QMS Conversion:
A Process Approach

QMS Conversion:
A Process Approach

David Hoyle

John Thompson

An imprint of Elsevier Science

Amsterdam London New York Oxford Paris Tokyo Boston San Diego San Francisco
Singapore Sydney

Butterworth-Heinemann is an imprint of Elsevier Science.

Copyright © 2002, Elsevier Science (USA). All rights reserved.

© Transition Support Ltd 2001   *Call*
Original Title: Converting a Quality Management System Using the Process Approach
Original ISBN: 1-903417-07-4

Library of Congress Cataloging-in-Publication Data

A catalog record for this book is available from the Library of Congress.

British Library Cataloging-in-Publication Data

A catalog record for this book is available from the British Library.

The publisher offers special discounts on bulk orders of this book.
For information, please contact:

Manager of Special Sales
Elsevier Science
225 Wildwood Avenue
Woburn, MA 01801-2041
Tel: 781-904-2500
Fax: 781-904-2620

For information on all Butterworth-Heinemann publications available, contact our World Wide Web home page at: http://www.bh.com.

10 9 8 7 6 5 4 3 2 1

Printed in the United States of America

About the authors

David Hoyle has over 30 years experience in quality management. He held managerial positions with British Aerospace and Ferranti International. As a management consultant—first, with Neville-Clarke Ltd and, before forming Transition Support Ltd, as an independent—he guided such companies as General Motors, Civil Aviation Authority and Bell Atlantic through their ISO 9000 programs. He has delivered quality management and auditor training courses throughout the world and has published five books with Butterworth Heinemann on ISO 9000, some of which have been translated into Spanish, Japanese, and Mandarin. Worldwide sales of his first book, now in its fourth edition, have totalled over 30,000 copies. He participates in various committees of the Institute of Quality Assurance and has been engaged in the revision of ISO 9000. He is a Chartered Engineer, a Fellow of the Institute of Quality Assurance, an IRCA registered Lead Auditor and a Member of the Royal Aeronautical Society.

John Thompson is an experienced management consultant in business improvement; over a 20-year period he has held management positions in Unilever, RHP Bearings, Mars and Caradon. During the past 12 years and prior to forming Transition Support Ltd, he was in management consultancy as a Director of Neville-Clarke Ltd and GPR Consultants Ltd. He assisted organizations in Europe, the Middle East, and Southeast Asia in their business improvement activities, including the use of ISO 9000 Baldrige, Singapore Quality Award and EFQM frameworks. He has helped many organizations to develop improvement strategies and apply the process approach to system development and to auditing. These included the Anchor Trust, Mars, TRW and MAFF. He is an adviser to the MTTA on its step change initiative. Initially trained as a statistician, he has undertaken post-graduate studies in business administration and is currently completing an MA in human resource management.

Contents

The past has only got us to where we are today

. . . it may not necessarily get us to where we want to be!

Foreword

Since 1987 the policies and practices of organizations that serve the achievement of quality have been inextricably linked with ISO 9000 that has become the most successful international standard ever.

Organizations were being told that all they had to do was "document what you do and do what you document." This simplistic and often misleading message spread so widely that probably over 90% of the quality systems developed since 1987, were merely collections of documents describing what organizations believed was needed to satisfy the 20 elements of the standard. This approach, the "element" or "clause" approach focused on conformity to requirements, often quite independently of business needs. As a consequence many organizations failed to realize any significant business benefit.

In the year 2000 revision of the ISO 9000 family of standards a fundamental change took place. The family of standards became based on eight principles of quality management that align well with the criteria in the EFQM Excellence Model. Completely rewritten and clearly focusing on customers, the "document what you do—do what you document" approach and the 20 elements were swept aside to be replaced with 5 sections that better reflect how organizations operate. But perhaps the most significant change was to promote a *process approach to quality management*. Requirements for documented procedures almost disappeared being replaced with requirements for processes. "Process" is not just another word for "procedure" but a totally different concept. The focus is on the mechanisms in an organization that enable it to satisfy customers. These mechanisms are the organization's "business processes" and this presents a significant opportunity for organizations to achieve real business benefits through application of ISO 9000:2000.

For those organizations that designed their systems and behaviors around the "old" 20 elements the 2000 revision created a significant change. It created the need for such quality management systems to be converted so that they used the process approach to quality management.

Written for executives, managers and practitioners, the purpose of this book is to help organizations see beyond the badge on the wall and gain benefits from their management system by converting their existing systems of documentation into systems of managed processes.

How to use this book

Each chapter in the book deals with a separate topic and each one has a set of learning outcomes that can be accomplished by covering its contents. The chapters follow the sequence of the conversion process. The first three chapters aim to change perceptions and establish the success criteria for the conversion. Chapter 1 provides the basis for understanding the approach taken by the ISO 9000:2000 family of standards. Chapter 2 aims to enable the reader to gain an understanding of the difference between procedures and processes. Chapter 3 introduces the eight quality management principles upon which the revision to ISO 9000 was based and illustrates the characteristics of an organization in which people are applying these principles.

Chapter 4 outlines the conversion process and explains the steps to be taken while Chapters 5 to 9 cover each of these steps in more detail. Chapter 8 is the most comprehensive in the book, explaining process analysis and the factors that need to be taken into account when designing and constructing effective processes. Chapter 10 addresses system validation which can be used at any stage in the conversion process as it contains criteria for judging a successful conversion and tracking progress.

Each chapter addresses the change in direction brought about by the ISO 9000:2000 family of standards. At the end of each chapter a summary is presented, often in terms of the differences between the old approach and the new approach, so as to continually emphasize the change in direction presented by the ISO 9000:2000 family of standards.

References to ISO 9000

Within this book reference to ISO 9000 without a year identifier refers to the old version of the family of standards unless a comparison is being made, when the form "ISO 9000:1994" is used to denote the old family and "ISO 9000:2000" to denote the new family of standards. Reference to the "2000 Standard" means the ISO 9001:2000. Reference to the "ISO 9000:2000 family" means

ISO 9000, ISO 9001, ISO 9004 and ISO 19011 and it is therefore recommended that as a minimum, both ISO 9000 and ISO 9004 should be studied as well as ISO 9001 in preparing for the conversion. "ISO 9000:2000" refers to the Fundamentals and vocabulary standard. "ISO 9001:2000" refers to the Requirements standard used for assessment and contractual purposes and "ISO 9004:2000" refers to the Guidelines for performance improvement. ISO 19011 refers to the Quality and environmental system audit standard planned for released in 2001.

Chapter 1

A real change in direction

Learning outcomes

After studying this chapter you should be able to:

- ❑ understand the real change in direction of the 2000 standard
- ❑ understand the reasons for change
- ❑ recognize the difference between conformance to standard and system performance
- ❑ explain the linkage between the role of the QMS and business outputs

A system focused on achieving business objectives

The ISO 9000:2000 family of standards is based on the process approach to management. This approach recognizes that all work is performed to achieve some objective—also that the objective is achieved more efficiently when related resources and activities are managed as a process. In addition it is believed that the objectives of the organization which serve to meet its mission will be met more effectively when the organization is managed as a system of interrelated processes.

It follows therefore that this system should be designed to enable the organization to meet its objectives and should interconnect all the processes required to deliver the desired results. Objectives are derived from the expectations of interested parties as now referred to in the ISO 9000:2000 family. Who are these interested parties? These for most organizations include:

- ❑ Customers who want products and services that fulfill their expectations

- ❑ Suppliers who want commercially viable and stable relationships

- ❑ Employees who want satisfying employment

- ❑ Shareholders who want a good return on their investment

- ❑ Society that wants organizations to operate responsibly, lawfully and ethically

None of these interested parties or stakeholders have objectives that are unrelated to the others and therefore they cannot have systems that operate independently—in fact there can only be one system. The process approach is therefore concerned with managing the interrelationships between the interested parties so that all are satisfied—not just customers. It is not a trade-off or a balancing act. Employee satisfaction or care for society or the environment cannot be traded-off against customer satisfaction. Clearly this is a change in focus and direction.

The fork in the road—old versus new interpretation

On first reading, the 2000 standard can be interpreted as shown in the Customer fulfillment cycle (Figure 1.1). Here there is a clear linkage between quality policy, objectives and where the QMS delivers the outputs to satisfy customer needs. To many this does not represent a significant change from how a QMS has been perceived. In reality the QMS was only a system of documentation focused on conformity to procedures (defined by a standard) as shown in the Conformity cycle (Figure 1.2). Here the linkage is between procedures which implement quality policy and deliver records that demonstrate conformity, often independent of business objectives. What the 2000

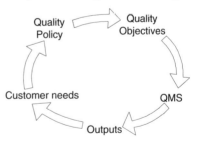

Figure 1.1 Customer fulfillment

standard requires is to go beyond conformity and seek customer satisfaction. However, because of the interdependencies referred to above, we have to go even further, beyond *customer* satisfaction and seek satisfaction of *all* the interested parties.

The 2000 standard needs to be interpreted as representing a QMS focused upon achieving business objectives as shown in the Business management cycle (Figure 1.3). Here the linkage is between a system of processes focused on achieving business objectives and satisfying the expectations of interested parties.

What is important to understand is the fundamental difference between Figures 1.2 and 1.3 and to interpret the changes to ISO 9000 as moving towards a new perception of the Business management cycle and not a continuance of the old perception as illustrated in Figure 1.4.

Figure 1.2 Conformity cycle

This view of a QMS is clearly a change in perception from a collection of procedures to the integration of business processes that involve people, technology, materials, equipment, facilities and the physical and human environment. However, as experience proves, it is the culture that most influences the effectiveness of such systems. No approach will therefore be successful without taking

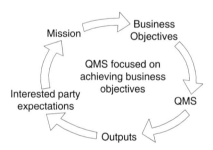

Figure 1.3 Business management cycle

full account of the prevailing culture in the organization. Whether the system is being developed from scratch or being converted from an established system, the development process is the same. The real differences emerge during system design, construction and operation. To convert a system that has been documented as a quality manual and a series of procedures firstly requires an understanding of the fundamental difference between procedures and processes and this is addressed in the next chapter.

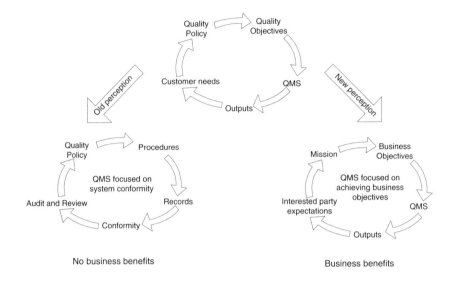

Figure 1.4 Convergence towards a business management cycle

What an ISO 9000:2000 organization looks like

An organization that has embraced the process management approach should be able to demonstrate a number of role model characteristics:

- ❑ A clearly defined business planning process that produces a robust business plan

- ❑ A business plan that consists of objectives, appropriate measures of success, actions focused on achieving those objectives with the relevant resources and skills provided

- ❑ An improvement culture and investment program to support continual improvement objectives

- ❑ Measured and monitored performance improvements in financial, environmental, quality, employee and customer satisfaction indicators

- ❑ Effective customer and market research processes linked to improvement planning

- ❏ Benchmarked performance against appropriate external data
- ❏ Awareness of position relative to competitors with known strengths and weaknesses
- ❏ Personnel development processes focused on releasing full potential
- ❏ Effective management of processes; i.e., processes that deliver outcomes which satisfy all the interested parties.

Summary

Clearly there is a real change in direction that results from comparing the old approach to ISO 9000, with its focus on conformity to standard, with the new approach with its focus on achieving real business benefits. The key differences in the two approaches are summarized in Table 1.1.

This chapter has highlighted the nature of the change in direction, in a way that should produce a change in perception about a QMS and give some insight into the magnitude of the challenge. In the next chapter, the difference between processes and procedures is explained in order to create an understanding essential for a successful conversion to be accomplished.

Table 1.1 Contrast between old and new approaches

Old approach	New approach
No clearly defined and communicated organizational purpose and objectives	Everyone understands the organization's purpose and objectives and is motivated and supported to achieve them
No marketing process and customer satisfaction measurement within QMS	Marketing process integrated in QMS and customer satisfaction regularly monitored
People are just another resource to be used to achieve the results	People are valued, developed and results achieved through team work
There is a set of random task based procedures that are independent of the business objectives	Processes are designed to achieve defined objectives and are continually measured, reviewed and improved
The system for achieving quality is defined by the 20 elements of ISO 9001:1994	Delivering business results is achieved through a coherent management system of integrated processes
Continual improvement is perceived as correcting mistakes only	Continual improvement is perceived as proactively seeking opportunities to improve performance at all levels and in all aspects
Data generated by the QMS that creates records is not used to make decisions	Decisions are based on performance data generated by the processes of the management system
Key decisions are made in an arbitrary and unilateral manner with purchasing decisions being based primarily on lowest price	Key decisions take into account the different stakeholders and their impact

Chapter 2

Processes versus procedures

Learning outcomes

After studying this chapter you should be able to:

- ❑ distinguish between procedures and processes

- ❑ understand what makes processes fundamentally different from procedures

- ❑ identify whether you have documented your procedures or your processes

- ❑ understand how big a gap exists between a procedural approach and a process approach

Change in direction

ISO 9001:1994 refers to documented procedures being required to implement almost every clause of the standard. In ISO 9001:2000, the requirement for documented procedures has been completely removed (apart from a few specific instances). In addition, the requirement for documentation has also been drastically reduced, signaling a change in direction away from documentation as the primary output from implementing ISO 9001:2000. The emphasis has moved from documented procedures to defined processes where the degree of documentation required is determined from an analysis of need. However, this change in direction as indicated in the Foreword and Chapter 1, is much more than a change in words. Procedures are not documented processes—as will become apparent from reading this chapter.

Procedures

Within the context of quality management standards, and more specifically ISO 9000, "procedure" was a key word that acquired a particular meaning over the years. In its simplest form a procedure is a way in which one works to accomplish a task. It can therefore be a sequence of steps that include preparation, implementation and completion of a task. Each step can be a sequence of activities and each activity a sequence of actions. The sequence of steps is critical to whether a statement or document is a procedure or something else. Specifications, contracts and records are not procedures as they do not tell us how to do anything. These describe the outputs resulting from carrying out procedures or tasks, leaving us to decide any further actions necessary to utilize these outputs. The output will more than likely be used as input to other procedures.

We need procedures when the task we have to perform is complex or when the task is routine and we want it to be performed consistently. If we are not concerned about how something is done and are interested only in the result, we do not produce procedures but issue instructions such as "post the letter," "repair the spin drier" or "recruit another person." These are work instructions as they intend us to do "quantitative" work without telling us how to do it or the "qualitative" standard to which the work should be carried out. Instructions are therefore not procedures unless they follow in a sequence and enable us to perform a task.

A set of self-assembly instructions is a procedure as it tells how to proceed to assemble the product. But the wording on the label telling us not to put hot objects on the surface is an instruction or a warning (a special type of instruction). As procedures are normally used by people they are designed with a user in mind. The user is normally an individual or a group of individuals, although procedures can cover a sequence of steps, each of which is performed by different individuals or groups. However, perceptions of procedures vary considerably depending on the context in which they are created and used.

Any sequence of steps, no matter how simple or complex can be expressed as a procedure that is intended to cause someone to take specific actions to accomplish a task. The key is that the steps follow a sequence. A random

8

collection of statements is not a procedure unless they are rearranged in a sequence that enables someone to proceed.

Misleading labels

Within the context of a QMS (and more specifically, ISO 9000) the procedure has, for many, taken on particular and sometimes peculiar characteristics. Such (documented) procedures may be written not as a sequence of activities or steps but as a series of requirements or a series of responsibilities. Neither of these can be procedures as they do not tell us how to proceed, what steps to take or how to measure the result.

Such procedures often follow a uniform format with a purpose statement, applicability statement, responsibilities and then procedure statements. Often there is no connection between the purpose statement and the procedure. Purpose statements often address the purpose of a document not the purpose of the task that the sequence of tasks is intended to deliver. Again, if such procedures (and rarely they do) contain measures of success, these are probably quantitative measures related to the task itself and not to why the procedure is carried out. The most common perception of such procedures is that they are simply associated with paperwork and filling in forms.

We seem to think that we have created a procedure by classifying a document as a procedure. These documents are often thought of as high-level procedures. Procedures do not have to be documented to be procedures and do not have to be high level. We often hear of procedures addressing the 20 elements of ISO 9000 and work instructions being used at departmental level to guide activities. These characteristics of procedures only serve to constrain our thoughts and our intent.

Processes

Processes convert inputs into outputs. They create a change of state. They take inputs (e.g., material, information, people) and pass these through a sequence of stages during which the inputs are transformed or their status changed to emerge as an output with different characteristics. Hence processes act upon inputs and are dormant until the input is received. At each stage the transformation tasks may be procedural, but may also be mechanical, chemical etc. Inherent processes do not normally recognize departmental or functional boundaries (but are often hindered by them) or the boundaries between customers and suppliers. Each process has an objective with both quantitative and qualitative measures of its outputs, directly related to its objectives. The transformation or process stages are designed to ensure the combination of resources achieves the objectives—the desired outputs. Of course this means that the process has to receive the right inputs to deliver the desired outputs and that the correct resources are applied at the right stages, in the correct quantities and in the right manner. It is true that a process can be illustrated as a sequence of steps just as a procedure is illustrated, but the similarity ends there.

Semantics

The way we use the words procedure and process tells us something about how they differ. We tend to start and stop processes. We implement procedures and commence and complete them. We process information. We do not procedure information but we may employ a procedure to process information. We have plating processes and there may be plating procedures. In this context, the plating process comprises the resources, people, plant and machinery, and the plating procedure contains the instructions on how to plate material. Clearly the instructions alone do not produce the results. Resources as well as behaviors impact outcomes and both need to be appropriate for any useful result to be produced.

We have process interrupt but not procedure interrupt, because processes are perceived as continuous and run until physical intervention. In our bodies we have processes, not procedures. The reproductive process, the digestive process, the respiratory process, these processes are certainly continuous and stop only when an intervention takes place. They may require human intervention in which a surgeon may employ procedures to effect a repair. Procedures on the other hand are perceived as being discontinuous, having steps which can be paused with activities or actions picked up or put down at will.

Procedures usually relate to groups of activities with a given output where that output may not be complete until acted upon by someone else at a later stage in the process. Therefore, procedures are the actions taken by individuals in a process that may span across several functions and use multiple resources to deliver a predetermined output at a given rate at a given location on a given date.

It should be clear by now that procedures are intended for use by people and are a product of the command and control age. It was believed that prescribing what people were required to do resulted in people doing the right things right first time. It was not appreciated that all work is a process and in addition to the commands, the required outputs will only be realized if the people have:

- ❑ The ability to do the job
- ❑ The motivation to do the job
- ❑ The resources to do the job

Where processes primarily differ from procedures is that processes are dynamic and procedures are static. Processes are dependent upon resources and the ability and motivation of the personnel involved to generate the desired outcome, whereas procedures are only a series of instructions for a person to follow.

The transition from procedures to processes is not simply changing labels on documents. Adding a flow chart does not turn a procedure into a process. Stringing together a series of statements as a chain of actions is not a process. A flow chart simply describes the sequential relationship between activities or tasks from start to finish and is not a practical method for showing how the activity is managed. Resources, human relationships and the behavioral factors that are present in every process need to be managed in addition to the sequence of tasks in order to deliver the desired outcomes. This transition is illustrated in Figure 2.1.

On the left side of the figure, the series of activities that act upon inputs to produce outputs is a procedure and take no account of the factors that cause results. On the right side the activities, resources and behaviors are represented as a process that is designed to achieve objectives. It receives inputs and generates outputs but the results are related to the process objectives, not simply whether outputs meet input requirements.

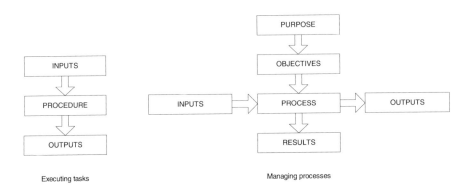

Figure 2.1 The transition from executing tasks to managing processes

Changing perceptions

Whether any change is necessary often depends upon one's current perception and attitude. If documented procedures merely respond to requirements in a standard they are not likely to demonstrate a clear line of sight to the real purpose of why the procedure is necessary. This is very often the root cause of why the people involved do not see the value in carrying out the actions. It is very likely that many organizations have to undergo a fairly radical rethink of the way they regard their management system. In order to satisfy the requirements, needs and expectations of the interested parties (customers, shareholders, employees, suppliers and the community) organizations have to identify the critical processes that deliver satisfaction. It is also clear that the effective management of those processes depends upon common understanding and monitoring of how success is measured. Perhaps the most crucial and radical factor is the constant focus and alignment of the key stages on achieving the end results.

The implication here is that in many cases the organization and the activities must have a greater alignment. The previous notion of a "functional" organization (for example, where different departments have created their own internal agendas, success criteria and cultures with no clear linkage to defined organizational objectives) has to be seriously questioned. It is all too tempting for organizations to address this issue by simply adopting a "cross-functional" approach which in reality means that they gather together representatives from different functional bunkers and let them fight it out. In general, a cross-functional team is not a substitute for process management. Simplistically, functions tend to be vertical while processes are horizontal, with stakeholder interest at both ends.

To make a transition from managing procedures to process management, an organization must establish whether it has:

❑ clearly defined what its objectives are and how it will measure and review the success of achieving these objectives

❑ evaluated the impact of those objectives on the interested parties—the stakeholders

❑ designed the critical, end-to-end processes necessary to deliver the objectives

❑ assessed and provided the resources, skills, competence and behaviors to make the processes work

The underlying principles of business excellence models and the changes to ISO 9000 should leave us in no doubt that the change in language from procedure to process is not about perception or semantics.

There is a very real change of focus required by many organizations if they are to remain competitive, and with this change comes the significant opportunity to ensure that their processes are designed to consistently add value.

Summary

In simple terms a procedure enables a task to be performed whereas a process enables a result to be achieved. This is the fundamental difference and should help the change in perception. Some other ways of perceiving procedures and processes that are discussed in this chapter are indicated in Table 2.1. It is only when the difference in concept has been understood that the whole perception of a QMS will change and that the change from procedure to process is not simply a change of words.

From reading Chapters 1 and 2, a clear understanding of the nature and magnitude of the change should now be forming. The reader should now appreciate why anyone approaching the conversion process should read these two chapters before proceeding further—to do otherwise runs the risk of

13

carrying the baggage of an obsolete approach into a new era and thus jeopardize the success of the conversion.

Table 2.1 Contrast between procedures and processes

Procedures	Processes
Procedures are driven by completion of the task	Processes are driven by achievement of a desired output
Procedures are implemented	Processes are operated and managed
Procedure steps may be completed by different people in different departments with different objectives	Process stages may be completed by different people but with the same objectives—departments do not matter
Procedures are discontinuous	Processes flow to conclusion
Procedures focus on satisfying the rules	Processes focus on satisfying the stakeholders
Procedures define the sequence of steps to execute a task	Processes transform inputs into outputs through use of resources
Procedures are driven by humans	Processes are driven by physical forces some of which may be human
Procedures may be used to process information	Information is processed by use of a procedure
Procedures exist—they are static	Processes behave—they are dynamic
Procedures cause people to take actions and decisions	Processes cause things to happen

Chapter 3

Quality management principles

Learning outcomes

After studying this chapter you should be able to:

- ❑ understand the importance and power of the eight quality management principles

- ❑ understand how the eight quality management principles relate to ISO 9001 and ISO 9004

- ❑ use the eight quality management principles to determine what has to change to create a QMS that meets the intent of ISO 9000:2000

- ❑ use the eight quality management principles in designing the management system

Change in direction

ISO 9001:2000 contains over 250 requirements and ISO 9004 over 268 recommendations—far too many to grasp what these standards aim to achieve. As a result many will treat the standards as a detailed action list and hence fail to realize the real benefits. This is like using legal statutes as a guide to life—there is much more than mere conformity. The standards were in fact based upon a set of eight quality management principles. By focusing on these principles we are able to see a much clearer picture of the real intent of the standard and the benefits to be gained. The principles enable us to determine the right things to do and understand why we are doing it. They provide a valid reason for using ISO 9000 beyond getting a "badge on the wall."

The eight quality management principles

A quality management principle is a comprehensive and fundamental rule or belief for assisting an organization in defining, identifying, guiding and validating organizational behaviors.

In June 1997, ISO/TC/176/SC2/WG15 produced a document (N130) that was to guide the development of a consistent pair of QA/QM standards. There is a strong relationship between these eight principles and the requirements of ISO 9001 and the guidance provided in ISO 9004. The principles are stated in ISO 9004 with an introduction that states, "Each major clause in this International Standard is based on the ISO 9000 eight quality management principles." In addition a joint communiqué from the IAF, ISO/TC-176 and ISO/CASCO issued in September 1999, required accreditation bodies to assess the certification body's transition programs to ensure that its auditors demonstrate knowledge and understanding of the eight Quality Management Principles. These principles and the implications for organizations with quality management systems that purport to meet the requirements of ISO 9000:1994 are explained below.

Customer focus

"Organizations depend on their customers and therefore should understand current and future customer needs, should meet customer requirements and strive to exceed customer expectations."

What does this mean?

> Customer focused means putting your energy into satisfying customers and understanding that profitability comes from satisfying customers. Expectations are created by the market place or a dominant supplier.

How does this impact the QMS?

> The QMS has not only to include the processing of current customer orders but has to include a process that identifies future market requirements. It also implies that a customer focus can be demonstrated throughout all processes within the system.

How will this be demonstrated?

> Demonstration of *Customer Focus* will be through the identification of customer-focused objectives throughout all processes within the system.

Leadership

"Leaders establish unity of purpose and direction. They should create and maintain the internal environment in which people can become fully involved in achieving the organization's objectives."

What does this mean?

> Leadership is providing role model behaviors consistent with the values of the organization—behaviors that will deliver the organization's objectives. The internal environment includes the culture and climate, management style, shared values, trust, motivation and support.

How does this impact the QMS?

> The manner in which the organization achieves its quality objectives becomes a key attribute of the QMS—i.e., culture & values are key to success.

How will this be demonstrated?

> Demonstration of *Leadership* will be through clearly developed and communicated vision and strategies and the regular review of performance.

Involvement of people

"People at all levels are the essence of an organization and their full involvement enables their abilities to be used for the organization's benefit."

What does this mean?

> Involving people means sharing knowledge, encouraging, developing and recognizing their contribution, utilizing their experience and operating with integrity.

How does this impact the QMS?

> The processes by which people are utilized, developed and treated together with the communication and sharing of knowledge become a key attribute of the QMS.

How will this be demonstrated?

> Demonstration of *Involvement of People* will be through discussions with staff regarding the setting of strategy, targets, the current and best ways of achieving objectives, and the feedback on their own and organizational performance.

Process approach

"A desired result is achieved more efficiently when activities and related resources are managed as a process."

What does this mean?

> Processes are dynamic—they cause things to happen. Procedures are static—they simply help people accomplish a task.

How does this impact the QMS?

> Work has to have a clearly defined purpose, be planned, implemented and the performance of the process measured, reviewed and improved.

How will this be demonstrated?

> Demonstration of the *Process Approach* will be through the definition of all the organization's business processes clearly

showing their contribution to achieving the organization's objectives and the mechanisms in place for managing them.

Systems approach to management

"Identifying, understanding and managing interrelated processes as a system contributes to the organization's effectiveness and efficiency in achieving its objectives."

What does this mean?

> Systems are constructed by connecting interrelated processes together to deliver the system objective that, in the case of the QMS, is the satisfaction of interested parties.

How does this impact the QMS?

> The QMS is not a random collection of elements, procedures and tasks but a set of interconnected processes that deliver the organization's objectives.

How will this be demonstrated?

> Demonstration of the *Systems Approach* will be through a definition of how all the organization's business processes interrelate with each other to achieve the organization's objectives and the evaluation of process changes for their impact on other processes and overall performance.

Continual improvement

"Continual improvement of the organization's overall performance should be a permanent objective of the organization."

What does this mean?

> Continual improvement is the progressive improvement in organizational efficiency and effectiveness—it is not about fixing problems; that is corrective action and part of process control.

How does this impact the QMS?

> Continual improvement becomes endemic such that everyone in the organization is actively seeking improvements in performance at all levels.

How will this be demonstrated?

> Demonstration of *Continual Improvement* will be through mechanisms for enabling staff to investigate current methods and performance in order to identify opportunities for improvement and that staff have the problem solving competence to deliver improved performance.

Factual approach to decision-making

"Effective decisions are based on the analysis of data and information."

What does this mean?

> Facts are obtained from observations performed by qualified people using qualified means of measurement—the integrity of the information is known.

How does this impact the QMS?

> Data required in making decisions affecting the business should be generated by the QMS and that all decisions can be substantiated by this data.

How will this be demonstrated?

> Demonstration of *Factual Approach* will be through a clear path to decisions from the analysis of information with confidence being created in the method of decision-making.

Mutually beneficial supplier relationships

"An organization and its suppliers are interdependent and a mutually beneficial relationship enhances the ability of both to create value."

What does this mean?

> Beneficial relationships are those in which both parties share knowledge, vision, values and understanding. Suppliers are not treated as adversaries.

How does this impact the QMS?

> The supply chain processes will be designed to foster involvement, co-operation, communication and a sense of partnership in achieving common and agreed objectives. There will be a move away from an adversarial approach to supplier relationships.

How will this be demonstrated?

> Demonstration of *Mutually Beneficial Supplier Relationships* will be through a supply chain process designed to foster involvement, co-operation, communication and a sense of partnership in achieving common and agreed objectives.

In Tables 3.1 to 3.8 the factors that characterize each principle are defined and the relationship shown between the principles and both ISO 9001 and ISO 9004. An explanation of the content in each column follows:

Application   These are the results one would expect to find in an organization that applied these principles. This has been based on the information provided in N130. However, entries marked * in Table 3.4 were not included in the original N130 document and have been added in order to recognize the feedback loop as a fundamental element of every process.

Motivation   These are the actions an organization would most likely have taken to achieve these results. These have been derived from ISO 9001 and ISO 9004.

ISO 9001 Clause   This is the clause number that most closely matches the result or the action. A blank indicates that the topic is not addressed in ISO 9001.
(All clauses match the principles)

ISO 9004 Clause   This is the clause number that most closely matches the result or the action. A blank indicates that ISO 9004 adds no more than contained in ISO 9001.

It is very significant that to address all the factors that characterize these principles, an organization would need to adopt both ISO 9001 and ISO 9004.

Conformity with ISO 9001 alone does not necessarily characterize an organization that embodies the eight quality management principles.

Table 3.1 Application of customer focus principle

Application An organization applying the customer focus principle would be one in which people:	Motivation The organization would have:	ISO 9001 Clause	ISO 9004 Clause
Understood customer needs and expectations	Established current and forecasted customer needs and expectations	5.2 7.2.1	5.1.2
	Translated customer needs and expectations into achievable requirements		5.2.2
	Identified statutory requirements that apply to its operations	5.2	5.2.3
	Periodically reviewed customer needs and expectations		7.2
	Periodically tested understanding of customer needs with the customer		7.2
Balanced the needs and expectations of all interested parties	Determined the needs of employees, suppliers, consumers, users, unions, partners, society		5.2.2
	Assessed the impact of solutions to customer needs on other interested parties		5.2.2
Communicated these needs and expectations throughout the organization	Communicated customer needs and expectations throughout the organization	5.1 5.5.2	5.2.2

Application An organization applying the customer focus principle would be one in which people:	Motivation The organization would have:	ISO 9001 Clause	ISO 9004 Clause
Have the knowledge, skills and resources required to satisfy the organization's customers	Linked customer needs and expectations with staff development programs		6.2.2.2
	Determined that the organization has the capability to meet commitments before acceptance	7.2.2	7.2
Measured customer satisfaction and acted on results	Measured customer satisfaction	8.2.1	5.1.1 8.2.1.2
Managed customer relationships	Made customers aware of capabilities and difficulties	7.2.3	
	Processed enquiries and orders effectively	7.2.3b	
	Processed all customer feedback	7.2.3c	
Could relate their goals and targets directly to customer needs and expectations	Established organizational goals and targets	5.4.1	5.1.1
	Deployed organizational goals and targets to each function and level in the organization		5.4.1
Acted upon the results of customer satisfaction measurements	Improved the performance of the organization to meet customer needs	8.5.1	8.2.1.2

Table 3.2 Application of leadership principle

Application An organization applying the leadership principle would be one in which leaders are:	Motivation The organization would have:	ISO 9001 Clause	ISO 9004 Clause
Being proactive and leading by example	Appointed competent managers who demonstrate role model behavior	5.1 6.2.1	5.1.2
Understanding and responding to changes in the external environment	A process for identifying and acting upon external changes that impact the organization		5.6.2 7.2
Considering the needs of all interested parties	Made a commitment to meeting customer needs and those of other interested parties	5.1	5.1.1
	Processes for managing the human factors of the work environment	6.4	6.4
Establishing a clear vision of the organization's future	A strategic planning process		5.1.1
	Established business policies consistent with its vision	5.3	5.3
Establishing shared values and ethical role models at all levels of the organization	Defined supporting values and behaviors for implementing its strategy and objectives		5.1.1
	Defined and demonstrable business ethics		5.2.3
Building trust and eliminating fear	Team building and development		6.2.2.2

Application An organization applying the leadership principle would be one in which leaders are:	Motivation The organization would have:	ISO 9001 Clause	ISO 9004 Clause
Providing people with the required resources and freedom to act with responsibility and accountability	Defined responsibility and authority of all personnel	5.5.1	5.5.1
	Processes for resourcing and empowering personnel	5.1d	8.3.1 8.5.4
Promoting open and honest communication	Internal communication, two-way reporting, staff briefings	5.5.3	5.5.3 6.2.1
Educating, training and coaching people	Staff appraisal, competence assessment and development processes	6.2.2	6.1.2 6.2.1 6.2.2.2
Setting challenging goals and targets	Management by objectives	5.4.1	5.4.1 5.4.2 8.5.4
Implementing strategy to achieve these goals and targets	Process management	4.1e	4.1 5.1.1 7.1

Table 3.3 Application of involvement of people principle

Application An organization applying the involvement of people principle would be one in which people are:	Motivation The organization would have:	ISO 9001 Clause	ISO 9004 Clause
Accepting ownership and responsibility to solve problems	Created conditions to encourage innovation		6.2.1
Actively seeking opportunities to make improvements	Quality improvement teams, corrective action teams, suggestion schemes		6.2.1 Annex B
Actively seeking opportunities to enhance their competencies, knowledge and experience	On-the-job training, facilitated workshops, assisted college education courses, external training, a library	6.2.2	6.1.2
Freely sharing knowledge and experience in teams and groups	Open intranet, discussion groups, quality improvement teams	5.5.3	7.1.2
Focusing on the creation of value for customers	Customer focus groups, suggestion schemes, rewards		6.2.1 8.5.4
Being innovative and creative in furthering the organization's objectives	Suggestion schemes, problems reporting processes, cost saving schemes, mistake proofing		6.2.1 7.6 8.5.4
Better representing the organization to customers, local communities and society at large	Staff at all levels liaising with customers	7.2.3	
	Staff involved in community projects, public speaking engagements, etc.		

Application An organization applying the involvement of people principle would be one in which people are:	Motivation The organization would have:	ISO 9001 Clause	ISO 9004 Clause
Deriving satisfaction from their work	Employee motivation process		5.1.2
	Employee satisfaction measurement		6.2.1
	Reward and recognition schemes		6.2.1 8.5.4
	Assigned jobs on the basis of competency	6.2.1	6.2.2.1
Enthusiastic and proud to be part of the organization	An environment in which people were recognized for their contribution		8.5.4

Table 3.4 Application of the process approach principle (see page 21)*

Application An organization applying the process approach principle would be one in which people are:	Motivation The organization would have:	ISO 9001 Clause	ISO 9004 Clause
Defining the process to achieve the desired result	Identified its business processes, determined their objectives and the activities and resources and the relationships required to accomplish them	4.1a	4.1 7.1.1 7.1.2
	Documented the business processes including those for product/service design, production, delivery, installation and servicing	7.3 7.4 7.5	4.2
Identifying and measuring the inputs and outputs of the process	Input and output measurement stages in each process	4.1e, 8.1 8.2.3	7.1.1 8.2.2
	Processes for measuring and monitoring of product	7.4.3, 8.2.4	8.2.3
Taking action to prevent use or delivery of nonconforming inputs or outputs until remedial action has been effected*	Processes to control nonconforming product	8.3	8.3
	Processes for dealing with customer complaints	7.2.3	8.2.1.2
Taking action to eliminate the cause of nonconforming inputs or outputs*	Corrective action processes for preventing the recurrence of problems	8.5.2	8.5.2

Application An organization applying the process approach principle would be one in which people are:	Motivation The organization would have:	ISO 9001 Clause	ISO 9004 Clause
Identifying the interfaces of the process with the functions of the organization	Process flow charts showing the functional responsibilities and relationships	4.1b	7.1.3.1
Evaluating possible risks, consequences and impacts of processes on customers, suppliers and other stakeholders of the process	Performed failure modes effect analysis	8.5.3	7.1.3.3 8.5.3
	Performed process capability studies	8.2.3	8.1.1
	Performed a cultural analysis		6.2.2.2
Establishing clear responsibility, authority, and accountability for managing the process	Appointed process owners and process operators		5.1.2
	Defined the responsibility and authority of those managing the process	5.5.1	
Identifying the internal and external customers, suppliers and other stakeholders of the process	Process descriptions that define process inputs and their source, the process outputs and their destination and the process outcomes and who is affected by them	4.2.2	5.1.2 7.1.2
	Process analysis tools being applied to all processes	4.1 6.1 6.2.2	7.1.3.2

Application An organization applying the process approach principle would be one in which people are:	Motivation The organization would have:	ISO 9001 Clause	ISO 9004 Clause
When designing processes, giving consideration to process steps, activities, flows, control measures, training needs, equipment, methods, information, materials and other resources to achieve the desired result	Determined the sequence and interaction of the processes	4.1b	7.1.3
	Determined the criteria and methods for ensuring effective operation of the processes	4.1c	7.1.3.2
	Provided information and resources necessary to support the processes	4.1d 6.1	7.1.3.1
	Established processes to control information used in and generated by the process	4.2.3	
	Provided the facilities and physical environment required to achieve the organization's objectives	6.3, 6.4	6.3, 6.4
	Established planning processes for processing specific products and services	7.1	7.1.3.1

Table 3.5 Application of the systems approach principle

Application An organization applying the system approach principle would be one in which people are:	Motivation The organization would have:	ISO 9001 Clause	ISO 9004 Clause
Defining the system by identifying or developing the processes that affect a given objective	Defined business objectives	5.4.1	5.4.1
	A management system of interconnected processes designed to achieve the business objectives	5.4.2	5.4.2
	Made available financial, human and physical resources to maintain and improve the system	6.1	6.7 6.8
	Documented the processes to the extent necessary for effective operation	4.1 4.2.2	4.2
	Personnel devoted to ensuring that the system has been established and maintained	5.5.2	5.5.2
Structuring the system to achieve the objective in the most efficient way	Established the system's inputs, outputs and interfaces		7.1.3.1
	Identified the key business processes and their sequence and interaction	4.1b	7.1.3.1
	Structured the system around the business processes		5.4.2

| Application | Motivation | | |
An organization applying the system approach principle would be one in which people are:	The organization would have:	ISO 9001 Clause	ISO 9004 Clause
Understanding the interdependencies among the processes of the system	Created a direct linkage between business results, the processes that deliver them and the objectives and policies of the organization that need to be achieved for it to accomplish its mission.	5.3 5.4.1 5.4.2 4.1a	5.1.2
Continually improving the system through measurement and evaluation	Created evaluation processes that feed information into improvement processes that cause improvement in system performance	8.2.2 8.5.1 5.6	8.5.4
	Created a continual improvement culture	5.3b	8.5.4
	Benchmarked processes and customer satisfaction		5.1.1 5.4.1 7.2 8.1.2
Establishing resource constraints prior to action	Designed all its processes by taking into account the resources and constraints that apply to each step within it	4.1d 6.1a	7.1.3.2
	Contingency plans that ensured business continuity in the event of a termination of natural resources		6.3 8.5.3

Table 3.6 Application of the continual improvement principle

Application An organization applying the continual improvement principles would be one in which people are:	Motivation The organization would have:	ISO 9001 Clause	ISO 9004 Clause
Making continual improvement of products, processes and systems an objective for every individual in the organization	Made a commitment to continually improve the quality management system	5.1 5.3b	5.1.2
	A staff appraisal process that evaluated performance on the basis of meeting continual improvement objectives		6.2.2.2
	Review and improvement mechanisms built into every process	4.1e	7.1.3.1
Applying the basic improvement concepts of incremental improvement and breakthrough improvement	Research teams that continually searched for innovative solutions to technological, managerial and social issues concerning the organization	8.5.1	5.6.2
	Improvement programs that identified major improvement projects to be undertaken	8.1	Annex B
	Continual process monitoring and measurement against improvement targets	8.2.3	5.1.1
	Established the capability of its processes and would be seeking ways to reduce variation	8.2.3	7.1.3.2

Application An organization applying the continual improvement principles would be one in which people are:	Motivation The organization would have:	ISO 9001 Clause	ISO 9004 Clause
Using periodic assessments against established criteria of excellence to identify areas for potential improvement	Established an internal audit process covering the whole organization	8.2.2	
	Conducted self-assessment to evaluate the organization against a recognized excellence model or similar criteria		8.2.1.5
Continually improving the efficiency and effectiveness of all processes	Continual improvement policies	5.3b	
	Appointed process owners assigned with the responsibility for improving process performance		5.1.2
	Acted upon results that identified opportunities for improvement	8.5.1	8.5.4
Promoting prevention-based activities	Established preventive action processes for all activities in the organization	8.5.3	8.5.3

Application	Motivation	ISO 9001	ISO 9004
An organization applying the continual improvement principles would be one in which people are:	The organization would have:	ISO 9001 Clause	ISO 9004 Clause
Providing every member of the organization with appropriate education and training, on the methods and tools of continual improvement	Established human resource development processes that included education and training in improvement methods		6.2.2.2
	Provided resources for implementing improvement programs	6.1	6.1.2
Establishing measures and goals to guide and track improvements	Established objectives and improvement targets for each process	5.4.1	
	Established mechanisms for measuring and monitoring process performance	8.2.3	
Recognizing improvements	Established reward and recognition schemes for employees achieving improvement goals		6.2.1 8.5.4

Table 3.7 Application of the factual approach principle

Application An organization applying the factual approach principle would be one in which people are:	Motivation The organization would have:	ISO 9001 Clause	ISO 9004 Clause
Taking measurements and collecting data and information relevant to the objective	Established objectives and targets for all processes	5.4.1 4.1c	8.1.1
	Established measurement and monitoring stages in all processes	4.1e 8.2.3	8.1.1
	Established processes for collecting and analyzing data from processes	8.4	8.4
	Established processes for measuring and monitoring product	8.2.4 7.4.3	8.2.3
	Personnel who collect and analyze the data	5.5.2	8.4
Ensuring the data and information are sufficiently accurate, reliable and accessible	Established processes for controlling recorded data	4.2.4	
	Established calibration processes for measuring device measurements	7.6	7.6
	Employed benchmarking techniques		7.2 8.1.2 8.2.1.5 8.4
	Used competitive analysis to judge validity of strategic decisions		5.2.2 8.4

Application An organization applying the factual approach principle would be one in which people are:	Motivation The organization would have:	ISO 9001 Clause	ISO 9004 Clause
Analyzing the data and information using valid methods	Validated the methods employed to analyze performance data	7.6	8.4
Understanding the value of appropriate statistical techniques	Established training programs in statistical methods	6.2.2	
	Established methods for identifying and applying appropriate statistical methods	8.1	8.1.2
Making decisions and taking action based on the results of logical analysis balance with experience and intuition	Identified the data needed for making decisions		8.1.1
	Established methods to convey validated performed data to decision makers		8.1.1
	Used the validated data when making decisions		8.1.1
	Established the capability of its processes	8.2.3	8.1.1

Table 3.8 Application of the mutual relationships principle

Application An organization applying the supplier relationship principle would be one in which people are:	Motivation The organization would have:	ISO 9001 Clause	ISO 9004 Clause
Identifying and selecting key suppliers	Established processes for supplier selection	7.4.1	7.4.2
Establishing supplier relationships that balance short-term gains with long-term considerations for the organization and society at large	Policies that treat suppliers as partners not adversaries		5.2.2
	Built sound relationships with its suppliers that are based on shared values		5.2.2
	Notified suppliers of its short- and long-term goals		5.2.2
Creating clear and open communications	Established channels of communication with suppliers that were not restrictive		5.2.2 6.6
	Established feedback mechanisms that provided suppliers with valid performance data on quality, delivery and cost		7.4.1

Application	Motivation	ISO 9001 Clause	ISO 9004 Clause
An organization applying the supplier relationship principle would be one in which people are:	The organization would have:		
Initiating joint development and improvement of products and processes	Set challenging goals and targets for suppliers		5.2.2
	Established supplier development programs for technology and quality management system improvement	8.5.1	6.6
Jointly establishing a clear understanding of customers' needs	Conveyed relevant customer needs and requirements to their suppliers		5.2.2
Sharing information and future plans	Established periodic reviews with suppliers that kept them informed of the organization's development		6.6 7.4.1
Recognizing supplier improvements and achievements	Established suppliers performance reviews		7.4.2
	Established award programs that recognized superior supplier performance		6.6

Using the principles to size the gap

The self-assessment tool that is presented in Table 3.9 will enable organizations to determine the size of the gap between where they are now and where they need to be to comply with ISO 9001:2000.

Table 3.9 Self-Assessment

CRITERIA	RATING
Customer focus 1-3 There is no proactive process for understanding customer needs. 4-6 We have a process but it is not in the QMS. 7-9 Our process is fully integrated within the QMS.	
Leadership 1-3 There is no clearly defined and communicated statement of the organization's purpose, values and objectives. 4-6 We know where we are going but we are not all pulling in the same direction. 7-9 Everyone understands the organization's purpose and objectives and is consistently supported to achieve them.	
Involvement of people 1-3 People are just another resource to be used to achieve our results. 4-6 We involve everyone in decisions that affect them. 7-9 We involve and value our people and achieve our results through teamwork.	
Process approach 1-3 The system is a set of random task-based procedures that is independent of the business objectives. 4-6 We have departmental processes that serve departmental goals. 7-9 We design our processes to meet defined objectives and continually measure, review and improve their performance	

CRITERIA	RATING
Systems approach 1-3 The system for achieving quality is organized around the 20 elements of ISO 9000. 4-6 We have formalized our operational processes so that they deliver conforming product. 7-9 We have integrated all our processes into a coherent management system that delivers the organization's objectives.	
Continual improvement 1-3 Continual improvement is perceived as correcting mistakes only. 4-6 Continual improvement is perceived as responding to problems. 7-9 Continual improvement is perceived as proactively seeking opportunities to improve performance.	
Factual approach 1-3 Data generated by the QMS is not used to make business decisions. 4-6 We mainly use audit data, customer complaints and nonconformity data as inputs to decision-making. 7-9 We base our decisions on process performance data generated by the management system.	
Mutually beneficial supplier relationships 1-3 Suppliers are treated as adversaries and kept at arm's length. 4-6 We work with our suppliers to improve our overall performance. 7-9 We involve our key suppliers in our future strategy.	
Total	

Interpretation of scores

A score of 8 to 24 is indicative of an organization that meets ISO 9001:1994, ISO 9002:1994 or ISO 9003:1994 but not ISO 9001:2000.

A score of 25 to 48 is indicative of an organization that has the potential for meeting ISO 9001:2000 but is not yet capable.

A score of 49 to 72 is indicative of an organization that meets the intent of ISO 9001:2000.

Using the principles to test process effectiveness

The principles can be used in validating the design of processes, in validating decisions, in auditing system and processes. You look at a process and ask:

❑ Where is the customer focus in this process?

❑ Where in this process is there leadership, guiding policies, measurable objectives and the environment that motivates the workforce to achieve these objectives?

❑ Where in this process is the involvement of people in the design of the process, the making of decisions, the monitoring and measurement of performance and the improvement of performance?

❑ Where in this process has the process approach been applied to the accomplishment of these objectives?

❑ Where in this process is the systems approach to the management of the interfacing processes, the optimization of performance, the elimination of bottlenecks?

❑ Where in this process are the facts collected and transmitted to the decision makers?

❑ Where in this process is there continual improvement in performance, efficiency and effectiveness?

❑ Where in this process is there a mutually beneficial relationship with suppliers?

Chapter 4

The conversion process

Learning outcomes

After studying this chapter you should be able to:

❑   understand the difference between the "document what you do" approach and the "process" approach to system development

❑   understand the steps to be taken to convert element-based systems to process-based systems

❑   identify the factors that affect success in the conversion process

❑   understand how application of the process approach to system conversion differs from the "document what you do" approach

Change in direction

ISO 9001:1994 required a quality system to be established, documented and maintained and the system and its documented procedures effectively implemented. ISO 9001:2000 requires a quality management system to be established, documented, implemented, maintained and continually improved. While the obvious difference is the inclusion of the requirement for continual improvement, the additional requirements on establishing the QMS go far beyond the notion that a QMS is a set of documents. There are requirements for the processes to be identified, their sequence and interaction determined, criteria and methods for their effective operation determined and process measurement, monitoring, analysis and improvement performed. These requirements signal a fundamental change in direction. They indicate that the "document what you do" approach would be an inappropriate approach to take. The QMS has to be designed to achieve planned results, i.e., business objectives—not simply conformity with procedures. The QMS is still required to be established and documented, but in view of what the QMS is now required to achieve we need to re-examine what "to establish" and "documented" really mean.

Previously, "to establish" was often interpreted as meaning "putting in place the documented procedures." In order to establish a "system" it has to be put in place and putting a "system" in place (rather than just a collection of procedures) requires several separate actions (assuming that we understand that a system is a collection of concepts, principles, resources, information, people and processes required to achieve defined objectives).

Previously "documented" was often interpreted as meaning that everything needed to be written down but this caused considerable difficulties in knowing what and how much to document. This was because of a conflict between the business need for documentation and the need to demonstrate conformity to external auditors. The ISO 9000:2000 family of standards recognizes that documentation should exist primarily to achieve business objectives.

The system firstly needs to be designed (to meet the defined objectives) after which it needs to be installed and commissioned before it can be considered operational. The approach is very similar to building any system, be it an air-conditioning system, a political system or an education system. However, the system is not in place until it has been integrated into the fabric of the organization so that people do the right things right without being told—then, and only then, can it be claimed the system is in place. This is a different concept, one which requires a new approach to building a QMS. Although many reading this book will work in organizations with existing systems (not necessarily established systems!) the process of converting an existing system is not simply replacing documented procedures with a set of flow charts, as will become apparent from reading this chapter.

Conversion process overview

The conversion of a QMS is a project with the objective of constructing a management system that enables an organization to achieve its objectives

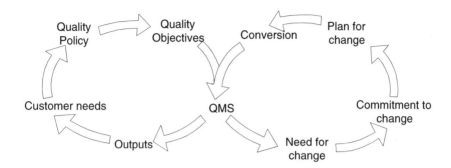

Figure 4.1 The conversion cycle

effectively and efficiently. The process commences with a need for change, which represents a gap between where we are and where we want to be. This is converted to a commitment to change when the climate is favorable and management is convinced of the need. A plan for change is developed and agreed for the conversion of the QMS into a system for achieving the "new objectives." The conversion itself is an iterative process during which perceptions of the system are changed, the business processes identified, captured and analyzed and changes introduced. Conversion is complete when the system delivers the desired results. This is illustrated in Figure 4.1 that shows in a figure-of-eight loop, how the current cycle is linked with the conversion cycle in a continual process of improvement.

The conversion process itself comprises two sub-processes—system design and system construction. Once the new policy and objectives have been defined, system design can commence. The system design process is complete when all the processes have been defined and a process development plan has been produced that shows how the system will be constructed.

The need for change

As stated in the Foreword, perhaps as many as 90% of the ISO 9000 registered organizations adopted the "element" or "clause" approach to QMS design and therefore for these organizations to achieve the business benefits from ISO 9001:2000, there is a compelling need for change. The change is about the way the QMS is perceived and managed, not necessarily about writing a completely new set of documents. Clearly, some new documents are needed but equally there may be many more old documents that are simply withdrawn, as they serve no useful purpose. For those with the task of convincing top management to invest in the change, the simple message is that the QMS must be focused on delivering the planned business outputs, not simply on maintaining certification to ISO 9000 (although that may have been their previous perception).

Commitment to change

The previous chapters provide the compelling reasons for change—what is needed is to present these in a manner that convinces top management to invest in the change. The first hurdle is to establish why a QMS is needed. While the initial reason may have been to obtain ISO 9000 certification because of customer pressure or marketing advantage, the purpose of a QMS is to enable the organization to meet its declared objectives. It should not be as was previously stated—to simply get ISO 9000 registration.

Part of the problem is the perception of the QMS. If the word Quality creates difficulties in winning the argument, don't use it. The system can still be compliant with ISO 9001:2000 if it is called a Management System, a Business Management System or anything else. It is not what it is called that matters—it is what it accomplishes and that all parties recognize that the system achieves business results and that they are part of it. However, for the purposes of the rest of this book we will refer to the converted system as BMS rather than QMS.

The second hurdle is getting buy-in from all parties concerned. Explaining that the BMS is about effective process management with the emphasis on business results and not about conformity to documented procedures should get total support.

Commitment is doing what you say you will do. However, it depends on the doers, knowing the

> A real change
>
> Quality
> Management
>
> to
>
> Business
> Management

right things to do and having a passion for the results they expect. In the past, one of the most common reasons why organizations encountered trouble 12-18 months after the initial ISO 9000 registration was a lack of commitment. They no longer did what they said they would do, but it was not because they were being deliberately dishonest—it was more likely they did not fully understand what the whole thing was about. Getting commitment is about creating a passion for the results and the road to commitment is often a six-stage process as shown in Table 4.1. You may need to lead people involved along this road before you have their commitment. Some people won't say things that they don't intend to do while others willingly say anything just to keep the peace!

Table 4.1. The road to commitment

Stage	Level	Meaning
0	Zero	I don't know anything about it.
1	Awareness	I know what it is and why I should do it.
2	Understanding	I know what I have to do and why I need to do it.
3	Investment	I have the resources to do it and I know how to deploy them.
4	Intent	This is what I am going to do and how I am going to do it.
5	Action	I have completed the first few actions and it has been successful.
6	Commitment	I am now doing everything I said I would do.

System design process

The System Design process is illustrated in Figure 4.2 indicating the tasks, inputs and outputs associated with its execution. Under the old "document what you do" approach, this phase would not have existed. Systems were not designed at all—whatever form the systems took—they were merely documented. The quality manual was a document that responded to the requirements of the standard and referenced the associated procedures, thus creating a system of documentation that mirrored the standard rather than the business. As the BMS is now perceived to comprise a series of interconnected processes that enable the organization to achieve its business objectives, it is now necessary to design a system to fulfill that specific purpose.

The System Design process commences with the development of a System Requirements Specification following which a model of the business is constructed in terms of its relationships with external interfaces and stakeholders. The core processes that deliver the organization's product should

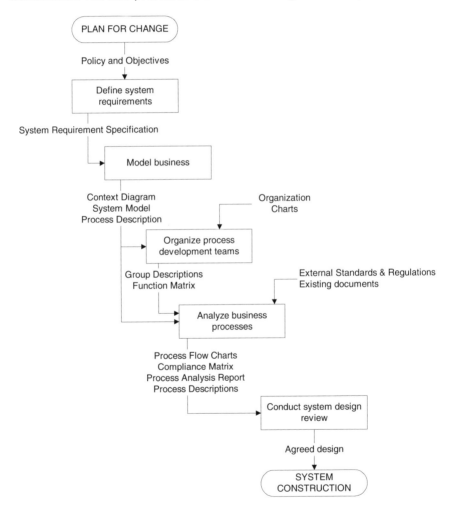

Figure 4.2 System design process

50

then be identified and these reviewed to ensure that no significant process has been overlooked.

Process Development Teams can be set up to carry out process analysis in order to identify the characteristics of each process. The primary output from the team is a suite of process flow charts that describe the sequence of task that transform business inputs into the desired business outputs. However these are not processes—they are merely descriptions of the sequence of activities.

The requirements of the external standards and regulations are deployed to the models in order to identify those current tasks that implement the requirements and those new tasks that need to be added to be compliant unless of course the specific requirements do not apply to the organization. Performance indicators and measurement methods are defined and the flow charts annotated to indicate the data collection points.

The results of the process analysis are documented and a series of process development plans prepared identifying all the changes required to enable the processes to meet the new objectives. The system design process ends when all the process models and development plans are agreed before proceeding to the system construction phase.

System construction

The system construction process is illustrated in Figure 4.3 indicating the principal stages that follow system design. Under the "document what you do" approach, this phase would cover the documentation and implementation stages. System construction commences with release of the system design consisting of the process descriptions, process analysis report and process development plan.

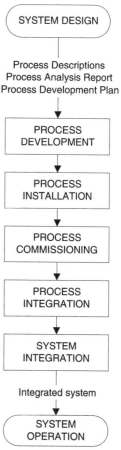

Figure 4.3 Construction process

The process descriptions contain details of the existing processes and the process analysis report identifies the information needs and recommended changes to be made to implement process management. The process development plan defines the stages through which the processes will be developed together with the responsibilities and timescales for their completion.

In the past, this process has been referred to as document development or simply implementation but these terms underestimate the work involved. Establishing an effective BMS means setting up a system on a permanent basis and putting it in place so that it functions effectively.

Following their design, the processes of the system need to be resourced, installed, commissioned and integrated into the fabric of the organization before it can be claimed that a system is in place. In order to construct the processes, information needs are matched with existing documentation and new information documented. Unlike the old "document what you do" approach, the information describing processes forms a coherent set so that all components of the system are described.

Process installation proceeds in parallel with process development so that new and revised practices are put in place at the same time as changes are being made in the cultural environment. The Installation stage involves decommissioning old procedures, preparing new foundations and installing new processes. The system needs to be constructed upon firm foundations and without the appropriate culture in place, the foundations will not be sound enough to enable processes to operate effectively. This was one of the weaknesses with the "document what you do" approach—it had little impact on the culture and hence did not create conditions for improved performance. Merely documenting current practice only formalizes it and runs the risk of making it more difficult to change if the documents do not reflect current practice accurately; the perception that its now "cast in concrete" often prevailed.

Following installation, processes are commissioned. This involves training the people, running or prototyping the processes, analyzing the results and fine-tuning them so that performance reaches the level required. During this stage, the monitoring components are activated, data is collected and performance evaluated.

After commissioning there follows an ongoing period during which the processes are integrated into the fabric of the organization. Process integration is about changing behavior so that people do the right things right because they know and believe it to be right. The steps within a process become routine, habits are formed and beliefs strengthened. The way people act and react to a certain stimulus becomes predictable and produces results that are predictable. System construction is complete when integration has taken place.

Summary

The conversion process is about changing the mind set from the "document what you do" approach to the process approach. The new approach treats the system as dynamic and integrated into the fabric of the organization and not a set of documents that are brought out just before the auditors arrive, dusted and updated. Differences in the approach are illustrated in Table 4.2.

This chapter has summarized the conversion process so as to provide an insight into what lies ahead. The next stage is to form a vision of where the organization wants to be and plan for the change as will be explained in the next chapter.

Table 4.2 Contrast between two approaches to QMS management

Document what you do approach	Process approach
Establish a QMS in response to customer pressure.	Establish a BMS as a means to accomplish the organization's goals.
Produce procedures as required by the standard.	Design a system of interconnected processes that reflects the operations of the business.
Produce a quality manual that responds to the standard.	Produce a system description that describes how business objectives are achieved.
Respond to the requirements of the standard.	Respond to the needs of the business.
Address the requirements of the standard in the order they are presented.	Deploy the requirements of the standard onto the business.
Document the procedures.	Document the processes.
Issue the procedures.	Commission processes.
Implement procedures.	Integrate processes into the fabric of the organization.

Chapter 5

Planning for change

Learning outcomes

After studying this chapter you should be able to:

- ❑ define the vision for the system in realistic terms
- ❑ understand how the quality management principles can be used to redefine the quality policy
- ❑ reassess the quality policy using the quality management principles
- ❑ define objectives for the organization that directly serve the mission statement
- ❑ establish the scope of the new BMS
- ❑ determine the characteristics of the BMS that will enable it to meet the defined objectives

Change in direction

ISO 9001:1994 required the policy for quality, including objectives for quality and the commitment to quality to be defined and documented. There was no criteria given that would guide users in producing an effective quality policy. Typical implementation of this requirement led to "motherhood" policies that were often identical from one organization to another. The quality policy was often something that management declared without understanding what it meant and what relationship it had to business policies. The objectives for quality were often interpreted as being included within the same policy statement and thus were not perceived as something that was measured. It

55

was not related to business objectives simply because the QMS was not perceived as central to business performance.

In the ISO 9000:2000 family there are five key changes here:

The meaning of quality—Quality now relates to the fulfillment of customers requirements as well as the requirements of other interested parties and not only *products that satisfy stated or implied needs*. This portrays a wider context that aligns well with organizational objectives and is a significant change.

Evidence of commitment—Top management are now required to provide evidence of commitment—not merely *define and document it*—but to clearly demonstrate commitment by providing vision and direction, communicating and involvement—another change in direction.

The scope of the quality policy—This is required to address continual improvement, customer requirements, regulatory and legal requirements and provide a framework for establishing the quality (organizational) objectives—another marked change in direction away from "motherhood" statements of little use.

Quality objectives—These are required to be measurable and consistent with the quality policy, within the wider context.

Processes and resources—Those needed to achieve the quality (organizational) objectives now have to be identified and planned—clearly a dramatic change in direction from requiring *documented procedures in accordance with the requirements of the standard*.

This clear link between policy, objectives and the processes to achieve them signals a fundamental change in direction that will be apparent from reading this chapter.

Defining system requirements

Few endeavors are successful without a clear vision of what is to be accomplished. Often the vision can be held in the mind but when several people are involved in achieving the vision it is wise to write it down. With a well-constructed specification addressing the important issues, everyone involved will be left in no doubt as to what the organization is attempting to do.

Following a decision to go ahead with developing a BMS or converting a QMS, a System Requirement Specification should be prepared that addresses the following points:

❏ Corporate policy or vision and values

❏ Business objectives, goals or aims

❏ System scope

❏ System design criteria or success factors for BMS design

These are addressed further below.

Corporate policy

Quality policies that were written within the context of the Conformity cycle (see Figure 1.2) were more likely to be focused on the system itself and not related to the business.

This policy need not be seen as a separate entity from the organization's stated business policies, which in some organizations are expressed as the vision, mission, values, strategy and intentions. In effect for simplicity, these expressions constitute the organization's mission as indicated in Figure 1.3.

Policies condition the behavior of those taking actions and decisions. They express the organization's values and intentions and set boundary conditions for what an organization will do or will not do and the manner in which it will do it. Policies therefore provide the guidance necessary for ensuring that actions and decisions are executed in a manner that will achieve the organization's objectives.

> Policies arise from asking the question:
>
> "What guiding principles do we intend to work by to accomplish our mission?"

The eight quality management principles referred to in ISO 9000:2000 and identified in Chapter 3, could be used to test the corporate policy. For example, how does the policy enable the organization to focus on its customers, how does it enable the involvement of people? Rather than repeating the words contained in the eight principles, business policies should be capable of demonstrating that they have been based on these principles. A corporate policy that does not include commitments to continually improve its performance is unlikely to be robust.

Business objectives

Quality objectives need not be seen as a separate entity from the organization's business objectives. In fact treating them as a separate entity often sets them apart such that they are perceived as being unrelated to business objectives. The business objectives should be derived directly from the mission statement and during the conversion process it is a good idea to capture these objectives in the System Requirement Specification so that they act as a constant reminder of the reason for the BMS.

There are two types of objectives: those that focus on results such as increasing sales, return on capital, profitability and those that focus on enablers that achieve these results such as, improving the competence of employees, reducing downtime of equipment, improving customer communication.

> Objectives arise from answering the question:
>
> "What factors affect our ability to accomplish our mission?"

The eight quality management principles represent key factors upon which business success depends. As explained in Chapter 3 few people can deny the critical importance of needing an acute awareness of stakeholder needs and expectations in order to establish and manage an organization to satisfy those needs. Ignoring the consumer, the environment, the employees, the suppliers or other interested parties including shareholders and trustees surely leads to disaster. Ignoring the need to design effective processes and manage these with relevant and factual information equally produces disaster. Therefore the eight principles could be used to define and test the business objectives. This could for example involve an assessment to test the presence of:

❑ Objectives for maintaining or improving customer focus relative to marketing, social responsibility and human resources

❑ Objectives for maintaining or improving leadership relative to marketing, innovation and human resources

❑ Objectives for maintaining or improving the involvement of people in all areas or specific areas

❑ Objectives for maintaining or improving the productivity of specific processes

❑ Objectives for maintaining or improving the processes by which continual improvement is accomplished in marketing, innovation, supply chain management

❑ Objectives for maintaining or improving the integrity of data used in making marketing decisions, innovation decisions, productivity decisions, etc.

❑ Objectives for maintaining or improving relationships with suppliers used in the marketing, innovation and production functions

In order to quantify these objectives the current performance in each of these areas needs to be established and targets chosen that are challenging but attainable. This assessment provides a useful input into establishing and validating the organization's critical success factors.

System scope

The scope of the system has in the past been related to what the organization required to be covered by its ISO 9000 certificate of registration. The very nature of the certification process allowed organizations to "cherry pick" the parts of their organization that were perceived as directly related to the standard and to select specific products and services for which the organization needed ISO 9000 certification.

This led to obvious anomalies; for example, design, servicing and distribution functions being excluded from the certificated system where clearly these functions were vital to satisfying customer requirements. It also led to multiple quality management systems—informal systems where ISO 9000 certification was not required, documented and audited systems where ISO 9000 certification was required and derivatives of these systems for other standards such as QS-9000, IAQS 9100, BS 8800 and ISO 14001.

In the 2000 standard, there are three dimensions to the scope of the system. The first is concerned with the extent to which the system covers all functions of the organization. The second is concerned with the extent to which the system covers all products and services provided by the organization and the third is concerned with the extent to which the system covers all requirements of the standard.

For those organizations that perceive the management system as the means by which its objectives are achieved, the BMS would cover all functions, all products and services and all relevant requirements of the standard. System scope is therefore not a question that needs to be addressed. However, according to ISO/TC 176/SC 2/N 524, an *organization is not obliged to include all the products that it provides within the scope of its system, or to address the realization processes for products that are not included within the system*. N524 also states that *where an organization finds that it is necessary to limit the application of the requirements of ISO 9001:2000, this must be defined and justified in the organization's Quality Manual*. No explanation is offered for why an organization would find it necessary to exclude certain functions, products or services from the system except to state that:

❏ *the system scope should be based on the nature of the organization's products and their realization processes, the result of risk assessment, commercial considerations, and contractual, statutory and regulatory requirements.*

❏ *claims of conformity to ISO 9001 are not acceptable unless these exclusions are limited to requirements within clause 7 and such exclusions do not affect the organization's ability, or responsibility, to provide a product that fulfills customer and applicable regulatory requirements*

❏ *any limitation in the system scope is required to be defined and justified in order to avoid confusing or misleading customers and end users.*

Therefore apart from there being circumstances when certain requirements of clause 7 would not apply, it would appear that there would be no justification for excluding any function, product, service or process.

A system that meets ISO 9001:2000 will embrace marketing, sales, finance, human resources, facility maintenance, distribution and any other activity that contributes towards the achievement of the business objectives. Therefore for most organizations, this will mean

> The scope of the system is the scope of the organization

a radical change as the scope of the QMS is redefined and moves towards a BMS in which the scope of the system is synonymous with the scope of the organization.

System design criteria

Clearly the BMS must enable the organization to achieve its objectives and the organization needs to determine what will affect its ability to do this. These are the factors that constitute a robust design. A business that relies on a rapid response must have a system that provides a rapid response not one that is lethargic. A high-tech business may need to constantly change its products and its organization, therefore it must have a system that is not constrained by who does what and where they fit in the organization.

To determine these design criteria, try completing this sentence for the current organization:

We need a system that . . .

Some examples are:

- We need a system that does not constrain us to the way we are currently organized. *We need freedom!*

- We need a system that encourages us to seek and use best practice. *We need innovation!*

- We need a system that is integrated into the fabric of the organization. *We need good habits!*

- We need a well-defined system that will drive us towards our goals. *We need support!*

- We need a system that enables us to anticipate events. *We need alertness!*

- ❏ We need a system to enable us to predict our performance. We need to be forward looking!

- ❏ We need a system that will prevent disruptions in business continuity. We need to identify potential risks!

- ❏ We need a system that enables us to control technological change and organizational change. We need to change management!

- ❏ We need a system that will give our customers confidence that we will meet their needs and expectations. We all need confidence!

Summary

Without a clear vision of what is to be accomplished no endeavor will be successful. Therefore in the conversion program it is important to set out with clearly defined goals and for everyone to understand the reason why the BMS is needed. The differences between the traditional approach and approach to quality policy and objectives are summarized in Table 5.1.

As a result of the consolidation of the three assessment standards (ISO 9001, 2 & 3 into ISO 9001:2000) the scope of the BMS can no longer omit key activities of the business. Organizations will now have to include all activities and this will mean a radical change as they redefine the scope of the old QMS.

Once the vision of where the organization wants to be has been established, work can commence on modeling the business as will be explained in the next chapter.

Table 5.1 Contrasting old and new approach to policy and objectives

Old approach	New approach
Define quality policy	Define the organization's mission
Ensure the quality policy looks like all others	Use the 8 QM principles to test the mission statement
Include quality objectives in quality policy	Derive measurable business objectives directly from the mission
Forget about quality objectives and focus on conformity with procedures	Use the 8 QM principles to test the business linkage and completeness of objectives
Include within the QMS only those functions that directly service customer orders	Include every process and hence every activity within the system—exclude none
No obvious connection between quality policy, objectives and procedures	Clear linkage between mission, objectives and processes

Chapter 6

Modeling the business

Learning outcomes

After studying this chapter you should be able to:

- ❏ construct a model of the business that identifies the key processes and their interfaces

- ❏ understand the nature of business processes and how they differ from functional activities

- ❏ understand the importance of the linkages between business processes and how to make the connections

- ❏ describe each of the business processes in terms that will facilitate their analysis and development

Change in direction

ISO 9001:1994 required a quality manual to be prepared covering the requirements of the standard and documented procedures to be prepared consistent with the requirements of the standard. This led to the approach of picking up the standard, paraphrasing the requirements in a manual and translating them directly into procedures. The result was a standard-led system that bore little relationship to the way business was conducted. The change in direction described in Chapter 5 clearly demonstrates that a new approach is now necessary—an approach that puts the business at the center and the standard in the role of providing a supporting framework rather than the driver in the center.

Who should do this?

Because the BMS has to be central to the business, top management are by default the system designers. It is therefore important that they are brought together as a team to share their perceptions of how the business is and should be managed to achieve its objectives. The management team should derive a common picture of the business. This picture can be represented by two models—a context diagram and system model.

The context diagram

Having defined a clear vision of what is to be accomplished, a diagram should be produced to place the organization in context with its stakeholders and their requirements and expectations. This can be referred to as a context diagram; an example of which is illustrated in Figure 6.1.

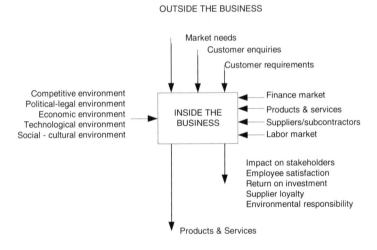

Figure 6.1 Context diagram

The interfaces should include three types of inputs:

a) the markets and customers that provide requirements and/or product for processing

b) external organizations that provide products, materials or information to accomplish the mission

c) internal organizations (but external to that part of the organization for which the quality system is being developed) that provide product or information to accomplish the organization's mission

The outputs should be represented as outcomes of the business in terms of:

a) products and services supplied (the physical outputs)

b) impacts on stakeholders (customers, employees, suppliers, investors, owners, society)

The result should be illustrated in a Context Diagram that shows where the organization fits in its environment. The specific product or information that passes along the channels that link the organization with its interfaces should also be specified.

These channels represent the principal channels of communication between the organization and its customers and other stakeholders. They provide the inputs and requirements for the result-producing processes within the organization.

The system model

From the Context Diagram a System Model should be developed that defines the core processes within the organization that convert the specified inputs into the required outputs. A System Model focused on the customer is shown in Figure 6.2. Other system models could be generated that focus on different stakeholders, e.g., employees, suppliers, shareholders.

These core processes are known as Business Processes. In general these are "end to end" series of activities that transform stakeholder needs into satisfaction. Everything that the organization does should fit into one or more of these processes so that they cover the organization's entire scope of work both operationally and administratively.

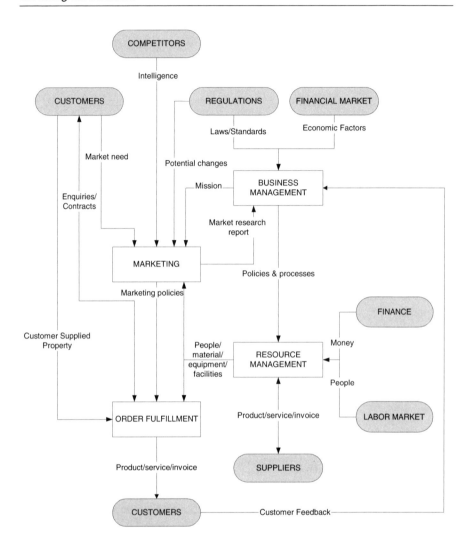

Figure 6.2 System model (customer focused)

The impact of the organization on its stakeholders is not shown in the model primarily because it results from the behavior of the participants and therefore emanates from every process.

The core processes in this generic model are described further below.

Business management process

The Business Management process (see Fig A.1) creates the vision, the mission and the overall strategy for the organization. Taking market research data from the marketing process, external standards, laws and other constraints, this strategy delivers the policies and objectives for the organization to fulfill in order to create and retain satisfied customers. It will determine the organizational configuration or design best suited to delivery of the strategy, for example, in-house or out-sourcing. It then develops the core processes needed to accomplish the organization's goals and deliver the desired results for all stakeholders. It is within the business management process that these business processes are developed creating a management system that serves business needs rather than a separate system created only to satisfy external auditors. The business management process develops and maintains a business planning cycle that regularly scans the external environment, monitors and reviews the overall performance against the organizational or external facing goals. In addition to this external perspective, it also reviews the performance of the business processes.

Marketing process

The marketing process (see Figure A.2) seeks to deliver new opportunities into the product/service realization process so that new products and services may be created to satisfy customer needs and expectations. This may include the key activities of market research, competitor analysis and promotion related to both current and future needs and expectations. The results of such research are translated into the organization's capability and potential customers directed into the sales process. The process predicts customer future needs which in the business management process are vital to forming the organization's strategy. The marketing process also maintains a vigilant view on the criteria that define customer satisfaction.

Order fulfillment process

The order fulfillment process (see Figure A.3) converts customer enquiries into satisfied customers and consists of three major sub-processes: sales, product/service realization and product distribution.

Sales process

The sales process (see Figure B.1) is the interface for contact with customers for existing products and services. It handles product enquiries, technical and expectation enquiries. It is the primary process for Enquiry Management and therefore plays a major role in knowledge about the customer. It processes contracts/orders and delivers requirements and knowledge into product/service generation processes or directly into the order-fulfillment process if product/service is already available for supply.

Product/Service realization process

The product/service realization process (see Figure B.2) transforms customer needs and expectations or specified requirements into products and services that satisfy customers. For existing product, this process replicates proven designs to consistent standards. For existing services, this process would involve the service delivery processes. For services with high tangible product content this may include maintenance and technical support. For services with low tangible product content, this may include consultations, health care, provision, investment services, etc. For new products/services, the requirements pass through a design process before emerging as a set of proven specifications that can be transformed into a tangible product/service. There are many variations within this process depending on the nature of the transaction between the organization and customer.

Product distribution process

The distribution process (see Figure B.3) supplies saleable product against customer requirements. In some cases the product may be in stock. In other cases the product may need to have been specifically produced or a service specifically designed to fulfill a given order. For tangible product, this process would include sub-processes of storage, packing, dispatch, shipment, installation and invoicing.

Where the organization's primary business is distribution, this process is synonymous with Product/service realization.

Resource management process

Triggered by the organization's objectives and requirements of specific customers, projects and initiatives, the resource management process (see Figure A.4) equips, maintains and develops the human, financial and physical resources needed to fulfill its objectives. It comprises several sub-processes concerned with resource planning, acquisition, deployment, maintenance and disposal. The resource would comprise people, products, materials, equipment, facilities as well as finance and hence involve, for example, purchasing, personnel, plant maintenance, financial and calibration services.

Naming the processes

The process names in Figure 6.2 are generic and not intended to equate with any particular function or department in an organization. The intention is to provide a useful starting point for discussion and development. Each organization should identify the business processes by names that appropriately describe the process. Where a proposed process name is the same as the name of an organizational function or department, an alternative name for the process should preferably be chosen. This will ensure that confusion is avoided and business processes remain multifunctional and not the perceived responsibility of a single function. Organizations are increasingly being more innovative in how they describe processes— "order to cash," "market opportunity to customer service offerings," "employee attraction and retention," "enquiry management." For our purpose we will continue to use "current" descriptors for processes, some of which we know may still be perceived by some as solely departments or functions. We do this to prevent confusion!

> Use process labels that reflect the purpose of the process not the name of the function

It is important to recognize that other roles and departments outside the sales and marketing departments participate in the sales and marketing process. The product/service generation process should be broken down into the key result-producing sub-processes such as product design, development, production, or service design etc.

Where there are separate departments for production and quality control for example, these should be depicted in terms of the processes executed not the functions that execute them. A functional relationship is illustrated in Figure 6.3 and a process relationship in Figure 6.4 indicating that the inspection and test activities of QC form sub-processes within the production process. When the production process is charted the separate tasks performed by the QC department would be identified.

71

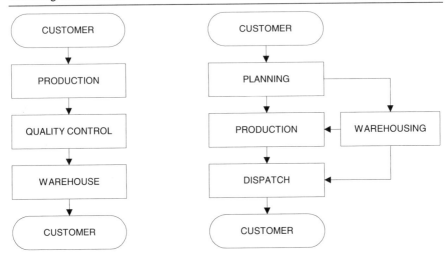

Figure 6.3 Functional relationship   Figure 6.4 Process relationship

Don't add every conceivable interconnection at this stage. Only the primary channels should be indicated.

Business process descriptions

On completion of the System Model, a description of each business process should be produced in order to unify understanding and agree on a common language. The process descriptions enable functions to determine which tasks belong to the process and those that do not. This may very well identify non-value adding tasks or activities. Process descriptions are not procedures or procedures containing a flow chart. The process descriptions should address the following:

a) The process objective (e.g., the objective of the resource management process is to determine, provide, effectively utilize, maintain and develop the resources required to achieve the organization's objectives.)

b) The indicators by which the achievement of the objective will be measured and the method of measurement

c) The process sponsor or owner (see Chapter 7)

d) Process inputs in terms of the materials and information to be processed

e) Process outputs in terms of the products, services and information delivered

f) The known factors upon which the achievement of the process output depends (e.g., skills and competence)

Previously, procedures were established as a direct response to the standard to meet a specific requirement e.g., contract review. A procedure was written and named "Contract Review." (Many organizations simply continued this "element" approach and wrote 19 more procedures, matching every element in the standard.) However contract review is a task that is part of the sales process the objectives of which are different. The real objective of the contract review process is to prevent the organization entering into a commitment with its customers that it is incapable of fulfilling, whereas the objective of the sales process is to secure business for the organization that meets the business objectives. Contract Review is therefore an activity designed to prevent failure whereas the Sales Process is a means to achieve the sales objectives. In reality, unfortunately, the output from most "contract review" procedures is a "tick in the box" or a signature indicating that the enquiry or order has been seen or acknowledged, not whether the organization has the capability to keep a promise to satisfy the customer needs and expectations!

Success indicators for many parts of the QMS were also "task" based, measuring whether the task had been carried out and the number of nonconformities found, not whether the real objective of the task had been accomplished. Therefore a success indicator for contract review was that "tick in a box" when it should have been the number of orders accepted which the organization knew it would meet in full.

Summary

The previous approach resulted in a model of the organization centered around the standard that bore little relationship to the way business was conducted (see Table 6.1). A more effective approach has been explained in this chapter that puts the business at the center and the standard in the role of providing a supporting framework for business effectiveness.

This has been shown as achievable by modeling the business as a series of "end to end" processes focused on achieving stakeholder objectives. These processes were shown to be those that transform stakeholder needs into satisfaction. This chapter has explained how the top management team can derive that common picture of the business and identify the key business processes that need to be in place to deliver the business results.

When the common picture or system model has been defined, teams can be formed to develop these business processes as will be explained in the next chapter.

Table 6.1 Contrast between old and new model of the business

Old approach	New approach
Quality manual that paraphrases the standard with little correlation to the business	System description that describes how the business is managed
Procedures that directly respond to and match the requirements of the standard	Business processes that achieve business objectives
Functionally based documentation	Process based documentation
Task based procedures	Objective driven processes
Focus on order driven activities	Focus on market and stakeholder driven processes

Chapter 7

Organizing process development teams

Learning outcomes

After studying this chapter you should be able to:

❑ understand the difference between the functional approach and the process approach to process development

❑ acknowledge that people will not instinctively think in terms of a process

❑ recognize that process development is an iterative process which constantly reveals new information about processes

❑ determine the relationships between functions and processes

Change in direction

ISO 9001:1994 required a quality manual and procedures consistent with the requirements of the standard and work instructions defining the manner of production, installation and servicing. This was often interpreted as three levels of documentation. Often an external consultant would write the level 1 quality manual, the quality manager or the consultant would write the 20 or so level 2 procedures and the functional managers or others would write the level 3 documents—the work instructions. Sometimes, one person with little or no involvement of others wrote the entire set of documentation. Although there was no requirement for manuals of work instructions for each function, it is how many organizations met the requirements. This led to a functional structure of documentation produced by people who, with few exceptions, did not meet with others outside their department in developing the documentation.

ISO 9001:2000 requires documented processes that achieve quality objectives. At the high level, such processes span many functions. It should be clear that each function cannot merely change its manuals of work instructions into manuals of process instructions because this will in itself not define the business processes.

All work is a process and is executed from beginning to end regardless of which department the people come from.

Many processes may be common to several departments. With the functional approach there is a tendency towards multiple ways of doing something rather than focusing on best practice for the organization. A change in direction is clearly required. A change that also creates the opportunity to streamline and reduce the amount of unnecessary documentation.

It is therefore necessary to bring the people together who contribute to a process in order to optimize effectiveness and how this can be accomplished will become apparent from reading this chapter.

Moving from function to process

As stated above, with the "document what you do" approach organizations often generated departmental manuals of procedures that only affected those within the department. There was often no reflection on why they were doing something as it was assumed that what they did actually generated what was required. There was also little consideration given to the interconnections between departments and their activities and the focus was on departmental conformity rather than on business performance. There was also a focus on departmental objectives often to the detriment of overall business performance. For example, purchasing departments often perceived they were meeting business objectives by minimizing their material spend, but in reality, the cheaper materials often caused excessive processing and rework further down stream. Finance departments often perceived their objective was to delay payment to suppliers in order to "save" money, only to lose production when the supplier stopped supply until payment resumed. This is symptomatic of the functional approach to QMS development. The functional approach structures

76

the QMS around who does what, and not around why things are done nor what they aim to achieve. A functional approach to setting objectives results in a functional approach being used for meeting them and monitoring their achievement and consequently, a function-oriented organization rather than a process-oriented organization.

The process approach starts with the overall objectives and works backwards, identifying the key activities required to achieve the objectives followed by the identification of skills, competencies and resources. With the process approach, the functions are secondary. The key to effective process development is to involve the people who contribute to the measured outputs.

The first step is to identify the activities that deliver the outputs then identify who performs the activities – initially this is not easy. Because people think functionally they often do not appreciate that they are part of a process that spans many functions. It is beneficial to ensure that people understand the nature of processes before involving them in identifying what processes are being used. Following identification of the high level business processes (see Chapter 6) the functions that carry out activities that contribute to each process can be identified. When representatives from the functions get together to identify process inputs, tasks, outputs, measures, etc., it will become apparent that other functions are involved but not present.

> **Ask which core business process people belong to, not which function.**

As more people are involved a clearer understanding of the process develops. We can refer to this grouping of people as the Process Development Team.

What may have been thought of as a process confined to a single function, emerges as a process with a "multifunctional dimension." It is also likely that as more thought is given to each process and its objectives, new interfaces and process boundaries emerge to make the original system model inadequate and further iteration necessary. It will also be revealed that some activities identified may not contribute effectively to any of the processes. These can be referred to as "non-value-adding activities" and may be subsequently dispensed with.

An example of this iterative process is the Human Resource Management Process. Initially, the only functions identified may be the HR Department. As the process development team analyzes the HR processes, it emerges that many other functions are involved in achieving HR process objectives and by the time the team reaches maturity, it may comprise representatives of every function in the organization. From ISO 9001:1994 the only procedure required for identifying training needs was very often simply satisfied by creating a training matrix and associated training records. From ISO 9001:2000 there are far wider requirements which can only be satisfied by defining the HR process in its entirety including HR planning, recruitment and selection or contracting, deployment and induction, training, development, welfare and termination. Also included must be the assessment of competence, effectiveness of training and evaluation of performance. There is such a vast difference from the 1994 version of the standard and therefore it is easy to see that the HR process is far more complex and multifunctional than the simplistic approach taken by ISO 9001:1994.

Deploying functions to model

There are a number of ways to establish an effective process development team. A well-tried method is to get together the functional heads to produce a Function Matrix that lists all Functional Groups, Departments, and Sections, etc., down the left side and the processes along the top. At the intersections a bullet is placed where the function contributes to the process. The bullet should represent key process actions or decisions taken by the function, not mere presence at a meeting or receipt of information. An example is shown in Table 7.1.

Creation of the Function Matrix is a key stage in identifying who is involved in what and helps involve the right people in the development process who will form a Process Development Team.

If some people are unsure, don't worry, as their role in each process will become clear once the process analysis begins. It is common, at this stage for people to realize, often for the first time, that everything they do is part of a process or processes. The concept of "joined up" is often taken for granted. It is only when the discussion and analysis take place that the concept becomes reality.

Table 7.1 Function matrix

Function	Business management	Marketing	Order fulfillment	Resource management
Executive Management	●	●		●
Marketing	●	●	●	●
Sales	●	●	●	●
Design	●	●	●	●
Purchasing	●		●	●
Production Planning	●		●	●
Production	●		●	●
Inspection	●		●	●
Inventory control	●		●	●
Maintenance	●			●
Quality Assurance	●	●	●	●
Dispatch	●		●	●

When each function has identified the processes to which it contributes, the identity of the person who will represent the function in a particular process should be determined. The Function Representative should be a person who is knowledgeable about the operations of the function and has authority to act on the group's behalf concerning the description of its practices.

When the representatives have been nominated, the bullets can be replaced with names. A person may represent more than one function if no other suitable person can be nominated. The Matrix then enables the personnel to be identified who will form the Process Development Teams.

Appointing the process "owner"

When the core business processes have been identified, a Process Owner or Sponsor should be appointed for each business process. The role of the process sponsor is to drive the process development effort and lead the team of people designing and constructing the processes so that they achieve the prescribed objectives. The process sponsor is someone who has the skills and competence to facilitate development and keep their eye on the ball. This may not necessarily be a functional manager. For processes that include several major sub-processes such as the Order Fulfillment Process, it may not be practical to appoint a specific individual as a process sponsor. An alternative solution is to set-up a Process Management Group comprising the process sponsors of all the sub-processes. Decisions affecting more than one process in the group would therefore be a group decision.

Team development

Each member of the process development team has to undergo process management training in which they understand the fundamental concepts of process management and the differences between the "document what you do" approach and the "process" approach. (Getting them to read Chapters 1 & 2 of this book would be a good start.) It is useful to start each team meeting with a reminder of process management concepts (see panel) to ensure that the development is kept on course.

Key concepts
All work is a process.
All processes serve as a means to achieve objectives.
Who does what is irrelevant providing they are competent to do it.
Every process has inputs, outputs, constraints and resources.
Every output must connect with another process and supply some of that process's inputs.
Every input must arise from the outputs of another process.
Every input, constraint or resource is supplied by a process.
Process management involves managing the behavior and interaction of events and measuring and monitoring their outputs.

Summary

The key to effective process development is to involve the people who contribute to the measured outputs. It is necessary to bring people together so that they can discover the real objectives of a process including who interfaces with whom, when and what passes between them. Process development is an iterative process. As more people are involved, a clearer understanding of the process develops. When process development teams have been formed, a more detailed process analysis can commence as will be explained in the next chapter.

Chapter 8

Process analysis

Learning outcomes

After studying this chapter you should be able to:

❑ describe the key elements of a process

❑ establish performance indicators and measurements for each process

❑ produce process flow charts that link together to form a coherent system description

❑ identify the extent to which the processes meet the requirements of governing standards

❑ identify the changes that need to be made to ensure the processes are effective

❑ select an appropriate type of document for communicating relevant process information

❑ understand the importance of culture on process performance and how to minimize its effects.

Change in direction

ISO 9001:1994 required production, installation and servicing processes to be identified (but no other processes). It also required these processes be planned and carried out under controlled conditions which were to include documented procedures and the monitoring of process parameters. ISO 9001:2000 takes a completely different approach. It requires the organization to measure, monitor and analyze processes, determine their sequence and interaction and determine criteria and methods to ensure effective operation and control. A process that is operating effectively delivers the required outputs of the required quality, on time, economically while meeting the policies and regulations that apply. This won't happen if left to chance, it has to be engineered—work has to be done to design a process with this understanding of effectiveness in mind.

Previously, all that was in most people's minds was to "document what you do." In some cases processes were operating effectively, but what was captured was at best a sequence of activities and at worst, a list of responsibilities. Much of what makes a process effective was left undiscovered, undocumented, not understood and probably not managed. What generally resulted from this approach was that independent procedures, not processes, were developed and written down and put together to form manuals. What is required is clearly a change in direction away from documenting what exists to designing effective processes—a task, the complexity of which will become apparent from reading this chapter.

Nature of processes

Processes comprise the actions and decisions required to transform the inputs into outputs that meet process objectives. However there are different types of activities and every activity requires adequate resources, information and a suitable environment for an effective transformation to take place. A popular way to define a process is through a flow chart. However, the flow chart should not be construed as being the process, as it is often merely a diagrammatic representation of the steps of a process.

To obtain a better understanding of the organization's processes it is necessary to perform a process analysis. Each process has a number of inherent characteristics.

- ❏ Products or information that are to be processed
- ❏ Objectives for the performance of the process
- ❏ Instructions which convey requirements for the product or information to be processed

- ❏ Planning activities which establish who is to do what, when, how, where and why

- ❏ Preparatory activities which set up conditions for commencing work

- ❏ Result-producing activities that act upon the inputs in the sequence they are executed

- ❏ Interfaces between activities and other processes supplying resources, product or information

- ❏ Interfaces between sequential activities receiving or supplying product or information required for processing

- ❏ Measurement activities for verifying that inputs and outputs meet requirements

- ❏ Measurement activities that verify that the process performs as intended

- ❏ Data collection points that capture data needed to judge process capability

- ❏ Diagnostic activities that discover the cause of variation

- ❏ Decision stages where decision makers consider the facts and decide on a course of action

- ❏ Feedback loops which return product or information for reprocessing

- ❏ Routing activities which move outputs including waste from one stage to another

- ❏ Resources which energize the activities and decisions including people, time, materials, machines, facilities, space, etc.

- ❏ Constraints that prevent, restrict, limit or regulate events.

Key process analysis tasks

The key tasks in the analysis of processes are listed below and described in more detail in subsequent sections. Information gathered about the existing processes should be recorded in a Process Description and the results of the analysis in a Process Analysis Report, as this will contain the changes to be made.

Process design

1. Determine the purpose and overall objectives of the process
2. Determine measures of success
3. Determine measurement methods
4. Identify key stages
5. Construct process flow charts for these key stages
6. For each stage:
    a. Define inputs
    b. Define resource requirements
    c. Define the tasks that make up the flow charts
    d. Define information needs
    e. Define the outputs and constraints, relating each to measures of success
7. Perform control analysis
    a. Identify input requirement
    b. Identify planning activities
    c. Identify implementation activities
    d. Identify checking/measuring/monitoring activities
    e. Identify feedback loops for remedial actions

    f.   Identify feedback loops for corrective actions

8. Conduct functional relationship analysis and simplify as necessary

9. Conduct analysis of relationships with other processes and simplify as necessary. This is key to ensure the provision and maintenance of information needs

10. Identify the specific skills and competencies required to deliver the outputs at each key stage

11. Identify other resource requirements

Process validation, and improvement

1. Determine current performance and future targets

2. Perform cultural analysis and prepare cultural change program as necessary

3. Perform failure modes and effects analysis and install failure prevention features as necessary

4. Perform productivity assessment and improve process efficiency as necessary

5. Deploy system requirements

    a.   Identify process element where requirement applies

    b.   Identify compliance or omissions

    c.   Insert additional elements if essential for process effectiveness

    d.   Complete system requirement compliance matrix

6. Deploy customer and other stakeholder requirements

    a.   Identify stages where requirement will be satisfied

    b.   Insert additional stages/elements to satisfy requirements

    c.   Complete stakeholder requirement compliance matrix

7. Repeat process performance measurement and recycle sequence as necessary

Key performance indicators (KPIs)

The process objectives were identified in Chapter 5 and having produced a diagrammatic representation of the process it is necessary to determine how achievement of these objectives will be indicated. For example, the objective of a purchasing process might be to supply the organization with the products and services it requires to meet its objectives. As the objectives may cover, financial, societal quality, productivity, innovation aspects, performance indicators for this process might be:

❑ Cost to the business of product/service under and oversupplied (right product, wrong quantity)

❑ Cost to the business of releasing defective product/service into the organization (right product, wrong quality)

❑ Number of occasions where the wrong product was released into the organization (wrong product)

❑ Cost to the business of purchasing from suppliers that fail to sustain delivery performance (right product, right quality, wrong time)

Performance measurement

For each performance indicator it is necessary to determine the method by which performance will be measured. The method will require:

❑ The data that needs to be collected

❑ Where the data is generated

❑ How the data will be collected

❑ How often the data needs to be collected

❑ What analysis needs to be carried out

❑ Who is to perform the data collection and analysis

❑ Who is to make the decision on whether performance meets target

Current performance assessment

In order to provide the driver for improvement the current performance from the process must be determined. Data should be collected using the agreed performance measures and the gap between desired and current performance determined. This gap gives the "Size of the Prize."

These process measures might be expressed in terms of:

a) Throughput (quantity of information or products that are processed in a given time)

b) Time through the process from receipt of inputs to release of outputs

c) Process start-up, set-up, shutdown or downtime

d) Operating costs

e) Satisfaction levels

It is vital to establish a clear linkage between the process objective and performance measures and the achievement of the business objectives.

Process flow charts

On completion of the System model, the descriptions of business processes and the Function Matrix, the Process Development Team can commence development of the process charts. The first series of charts to develop is the business process charts. As each is completed, charts for each work process can be developed. This is an iterative process whereby the business process charts and system model are revisited to modify them as more and more facts emerge.

Some simple charting conventions are shown in Figure 8.1. These are not drawing logic charts or software data flow charts—these represent the sequence of actions and decisions that occur in the process and notes of some key facts about them.

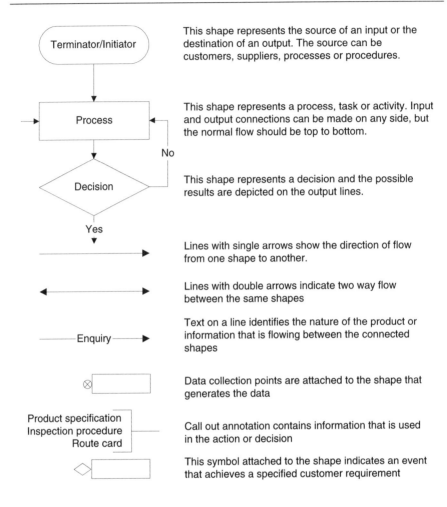

Figure 8.1 Charting conventions

Taking the inputs and outputs defined in the system model; describe each business process in flow chart form. Draw the process flow chart showing:

a) Interfaces with other processes (the source of the inputs and destination of the outputs)

b) Interfaces with external organizations

c) Sequence of tasks (what happens next)

d) Responsibilities (who does what)

e) Documentation used

f) Data recording points

g) Data collection points (data for measuring and deciding process/product performance)

h) Routing of collected data for analysis

i) The stages where specified customer requirements are achieved.

The chart should reflect how business is currently conducted, not how people think it is conducted or how it should be conducted. An example is shown in Figure 8.2.

Although, it is not the purpose of quality system development to merely document what you currently do, it is necessary to commence with an accurate description of processes, as they currently exist. Chart the sequence of events that follow receipt of the input from the feeding processes to the output that is transmitted to the receiving processes. Either start or end with a link to that process. All work requires an input to commence. In many cases the input is either a product or a piece of information. Even with time dependent tasks, the task commences with the release of the time schedule that stipulates when the task is to be performed. Ask the question "Where do the instructions come from to trigger this process?" The decomposition of the system is illustrated in Figure 8.3. The diagram shows alternative nomenclature. The sub-system category may be used to classify major processes that comprise processes or sub-processes. Product realization and resource management are examples where an intermediate division of processes is needed in order to capture different versions of the same generic process. For instance, the process for acquiring physical resources will differ from the process for acquiring human resources.

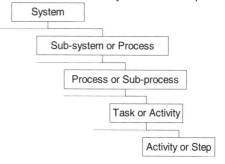

Figure 8.3 System decomposition

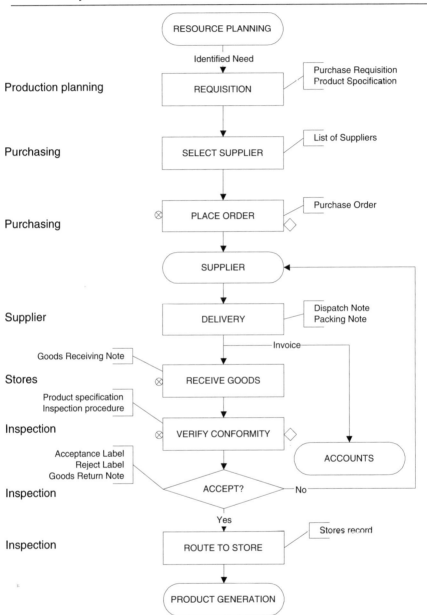

Figure 8.2 Sample process flow chart

The decomposition of processes is illustrated in Figure 8.4. Here we take one of the business processes of the system model, identify the tasks that constitute the business process and form the second tier of flow charts. Next we take each task of a business process and identify the activities that constitute the work processes to form further flow charts.

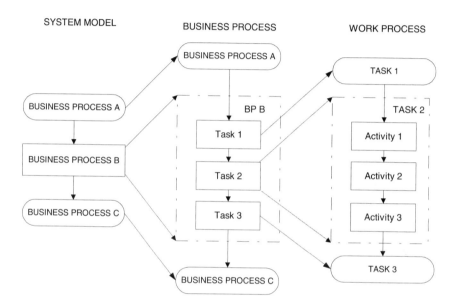

Figure 8.4 Process decomposition

The inputs and outputs should be shown at each stage in the decomposition. This illustration is symbolic. The tasks may interface with other processes and there may be decision tasks as well as action tasks.

The tasks in the business process chart are likely to cross-functional boundaries so that the responsibility for each task is different. At the work process level, the responsibility for each activity may be the same but it may take several tiers before reaching this level.

So that the set of charts reflects a coherent system, it is important to ensure all process linkages are in place. Ensure all process initiators and terminators link to other processes, or in the case of business inputs and outputs, ensure they link to the external organizations. Also ensure the titles used in the initiators and terminators are the same as the processes feeding the inputs or receiving the outputs.

Task analysis

Often what is known about a task is that which is described in the related procedures. If these have been produced using ISO 9001:1994 or ISO 9002:1994 as the guide, many aspects essential for its effective operation may have been overlooked. The task itself may also not fulfill a useful purpose within the process.

For each task in the flow chart ask:

- ❏ Is the objective of the task clearly understood?
- ❏ Does the task provide an output which serves the process objective?
- ❏ Do we have measures for its performance?
- ❏ Have we got an effective method of measuring performance?
- ❏ Do we have results that indicate that the task is performing to target?
- ❏ Is the output provided at the appropriate stage in the process?
- ❏ Is this the only process that generates this output?

If the answer to any of these is "no" the issue needs to be resolved before proceeding further.

It is also useful to perform a comprehensive task analysis in order to discover the characteristics upon which its effectiveness depends. Such an analysis will reveal the current basis for performing the task as well as any omissions. A useful technique is to analyze each task on the flow chart by applying *Kipling's Law* which states:

"I keep six honest serving-men

They taught me all I know;

Their names are What and Why and When

And How and Where and Who."

(*Rudyard Kipling*).

94

Table 8.1 Task analysis questionnaire

Inputs	• What are the inputs? • Why are they supplied? • Where do they come from? • Who supplies them? • How are they supplied? • When are they supplied?	• How are they measured? • What happens if inputs are incorrect? • What source traceability is required? • How frequently are they supplied? • How are they held pending use? • What happens if they are late?
Resources	• What resources are required? • Why are they needed? • Where do they come from? • Who supplies them? • How are they supplied? • When are they supplied?	• What skills and competencies are required to deliver the outputs? • What information is required? • How frequently are they required? • How are they measured? • What happens if they are incorrect? • What source traceability is required?
Task	• What is performed? • Why is it performed? • Where is it performed? • Who performs it? • How is it performed? • When is it performed?	• What is the preceding step? • What approvals are needed to start? • What precautions need to be observed? • How is performance measured? • What happens when failure is encountered? • What happens if the task is omitted?
Constraints	• What are the constraints? • Why are the constraints necessary? • Where in this process are they applied? • Who imposes the constraints? • How are they addressed? • When do they apply?	• What effect do they have on performance? • Who checks that the constraints are maintained? • What happens if they are not maintained? • How frequently are the checks performed? • What is at risk if constraints are removed? • How would their removal be detected?
Outputs	• What are the outputs? • Why are they needed? • Where do they go? • Who receives them? • How are they supplied? • When are they supplied?	• How are they measured? • What happens if they are not correct? • What source traceability is required? • What happens if they are late? • How is waste dealt with? • What happens next?

Table 8.1 illustrates how this can be put into practice to obtain information that uniquely characterizes each task. The results may be used as inputs into a Process Description or displayed on the relevant process charts.

Process constraints

The constraints on a process are the things that limit its freedom. Actions should be performed within the boundaries of the law, regulations impose conditions on hygiene, emissions and the internal and external environment. They may constrain resources (including time and costs), effects, methods, decisions and many other factors depending on the type of process, the risks and its significance with respect to the business and society. Almost all constraints are imposed by stakeholders. Therefore constraints invariably are related to outcomes or measurable outputs.

Process resources

The resources in a process are the supplies that can be drawn upon when needed by the process. Resources are classified into human, physical, information and financial resources. The human resources include managers and staff including employees, contractors, volunteers and partners. Physical resources include materials, equipment, plans, machinery but also include time. Information includes all the necessary knowledge required to enable the human and physical resources to achieve the process outputs. The financial resources include money, credit and sponsorship. Resources are used or consumed by the process. People are a resource when used by the process but are inputs when transformed by a process such as a training process that imparts skills.

Each process manages the utilization of resources not their acquisition, deployment or maintenance. The resources required by a process are therefore acquired and delivered to the process from a resource management process. This resource management process also plans, maintains and disposes of these resources.

All processes require physical resources that are available and capable, human resources that are available and competent, financial resources that are available and sufficient and information resources that are available and dependable.

Capability

When assessing physical resources it is necessary to determine the capability required to deliver the process objectives. For example, production equipment would need to be serviceable, safe and capable of producing the product features required within specified limits. Measuring equipment would need to be serviceable, safe and capable of measuring the required parameters accurately and precisely (i.e., calibrated). Lifting equipment would also need to be serviceable, safe and capable of carrying the load required.

Competence

When assessing human resources it is necessary to determine the competence required to deliver the process objectives. For example, operations personnel would need to demonstrate the ability to:

- ❑ Understand and interpret technical specifications

- ❑ Set up equipment

- ❑ Operate the equipment so as to produce the required output

- ❑ Undertake accurate measurements

- ❑ Apply variation theory to the identification of problems

- ❑ Apply problem-solving methods to maintain control of the process.

Simply possessing the ability to operate a machine is not a mark of competence. Competence is concerned with demonstrating the ability to achieve objectives. The competent operator delivers results that meet the process objectives. When identifying the competence needed, it is important to distinguish between competence and qualification. If a person has the appropriate education, training and skills to perform a job the person can be considered qualified. If a person demonstrates the ability to achieve the desired results the person can be considered competent.

However:

❑ A person may have received training but not have had the opportunity to apply the knowledge and skills required to achieve the results, therefore is "not yet competent."

❑ A person may have practiced the acquired skills but not reached a level of proficiency to work unsupervised. Again, "not yet competent."

❑ A person may possess the knowledge and skills required for a job but may be temporarily or permanently incapacitated. (A professional football player with a broken leg is not competent to play the game until his leg has healed and he is back in top form). Again, sadly, "not yet competent"!

Dependability

When assessing information resources it is necessary to establish their integrity, accuracy, legitimacy, authenticity and hence their dependability. Information from an unreliable source is likely to be suspect. Information that has not completed the preparation cycle and passed through relevant approval stages may be incomplete or inaccurate. Information that was produced using invalid data is likely to be unsuitable. Information that is simply not relevant or obsolete is likely to be inadequate. Information needs to be deemed dependable before use in the process. It is therefore the function of the process that produces information to ensure its dependability before being routed to processes that use the information. The process using the information simply verifies that it is received from a dependable source.

Control analysis

Once the initial chart has been developed a control analysis should be performed to identify key controls. Using the generic process model in Figure 8.5, examine the initial process charts to identify the following:

a) What constitutes the input requirement and how is this measured?

b) What constitutes the planning or preparation tasks? (i.e., What has to be in place to start the process correctly?)

c) What constitutes the doing tasks? (i.e., What has to be done to produce the correct outputs?)

d) What constitutes the checking tasks before release of process output? (i.e., When and how are the outputs measured?)

e) What constitutes the remedial action tasks in the event that the checking tasks reveal problems?

f) What constitutes the corrective action tasks?

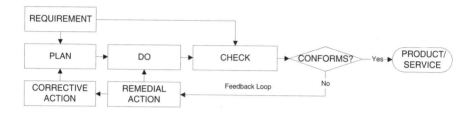

Figure 8.5 Generic control process

If a) to f) above cannot be satisfied and the equivalent tasks identified in the process chart, add tasks or redefine tasks to indicate both product and process controls.

The inputs should be shown on the input feed lines and outputs on the output feed lines. The interfaces between processes should be matched with one another so that the charts represent the flow of work through a process and the complete suite of charts reflects a coherent system with no gaps, overlaps, dead ends or loose ends.

System requirement deployment

An approach taken to ISO 9000 by many organizations was to respond to the requirements of the standard in the sequence they were stated. This approach assumed that the requirements were firstly listed in a logical order and secondly that the requirements covered every process the organization needed to employ.

The requirements of ISO 9001:1994 were those considered necessary to provide an adequate assurance of product quality and not those provisions that were necessary to market, design, procure, produce, supply and maintain products that satisfy customers. There are therefore activities that were not addressed by the standard.

A more appropriate approach is to model the business as described in Chapter 6 and then deploy or map the requirements onto the identified processes. In this way any gaps will be identified where the processes would be noncompliant.

After creating the system model, associated business processes and work process charts, the requirements of the external standards and regulations should be deployed onto the flow charts and a series of Compliance Matrices of the form in Table 8.2 produced. An even more robust approach is to use ISO 9004 in the matrix thereby identifying opportunities for improvement beyond mere compliance with ISO 9001:2000.

Where requirements of the standard cannot be matched with any task or decision in the set of charts, a gap may have been identified. In principle, a requirement of an external standard (e.g., ISO 9001, ISO/TS 16949, QS-9000) or regulation (e.g., Environmental Pollution Act) would only result in a new task or activity on a process chart if the requirement is related to a task or activity not currently carried out in the organization.

Table 8.2 Compliance matrix

ISO 9001 Clause	Title	Business management	Marketing	Order fulfillment	Resource management
4.1	General requirements	●			
4.2.1	General documentation requirements	●			
4.2.2	Quality manual	●			
4.2.3	Control of documents	●			
4.2.4	Control of quality records	●			
5.1	Management commitment	●			
5.2	Customer focus	●	●	●	●
5.3	Quality policy	●			
5.4.1	Quality objectives	●	●	●	●
5.4.2	Quality management system planning	●	●	●	●
5.5.1	Responsibility and authority	●	●	●	●
5.5.2	Management representative	●			
5.5.3	Internal communication	●	●	●	●
5.6.1	Management review—general	●			

ISO 9001 Clause	Title	Business management	Marketing	Order fulfillment	Resource management
5.6.2	Review input	●	●	●	●
5.6.3	Review output	●			
6.1	Provision of resources	●	●	●	●
6.2.1	Human resources—general	●	●	●	●
6.2.2	Competency, awareness and training				●
6.3	Infrastructure				●
6.4	Work environment	●	●	●	●
7.1	Planning of realization processes	●	●	●	●
7.2.1	Determination of requirements related to the product		●	●	
7.2.2	Review of requirements related to the product			●	
7.2.3	Customer communication	●	●	●	
7.3.1	Design and development planning			●	
7.3.2	Design and development inputs			●	
7.3.3	Design and development outputs			●	

ISO 9001 Clause	Title	Business management	Marketing	Order fulfillment	Resource management
7.3.4	Design and development review			●	
7.3.5	Design and/or development verification			●	
7.3.6	Design and development validation			●	
7.3.7	Control of design and development changes			●	
7.4.1	Purchasing process			●	●
7.4.2	Purchasing information				●
7.4.3	Verification of purchased product				●
7.5	Production and service provision			●	
7.5.1	Control of production and service provision			●	
7.5.2	Validation of processes			●	
7.5.3	Identification and traceability			●	●
7.5.4	Customer property			●	●
7.5.5	Preservation of product			●	●
7.6	Control of measuring and monitoring devices			●	●

ISO 9001 Clause	Title	Business management	Marketing	Order fulfillment	Resource management
8.1	Measurement, analysis and improvement—general	●	●	●	●
8.2.1	Customer satisfaction		●	●	
8.2.2	Internal audit	●	●	●	●
8.2.3	Measurement and monitoring of processes	●	●	●	●
8.2.4	Measurement and monitoring of product			●	
8.3	Control of nonconformity			●	
8.4	Analysis of data	●	●	●	●
8.5.1	Continual improvement	●	●	●	●
8.5.2	Corrective action	●	●	●	●
8.5.3	Preventive action	●	●	●	●

Customer requirement deployment

Process stages where specific customer needs and expectations are satisfied should be identified and the chart annotated with the designated symbol (see Figure 8.1).

Failure modes and effects

A powerful but under-utilized method of improving the effectiveness and performance of processes is Failure Modes and Effects Analysis (FMEA). A failure modes and effects analysis is a systematic analytical technique for identifying potential failures in the design of a product or process, assessing the probability of occurrence and likely effect and determining the measures needed to eliminate, contain or control the effects.

Firstly each function of the process is identified, i.e., an outcome the process is designed to accomplish. For each outcome it is necessary to establish the:

a) Potential failure modes (the manner in which the process could potentially fail to meet the design intent)

b) Potential effects of failure in terms of what the customer of the output might notice or experience

c) Severity of the effect (a numerical range of 1 to 10, can be used to grade severity with 10 being hazardous without warning and 1 having no effect)

d) Classification of the effect in terms of whether it is critical, key, major or significant

e) Potential cause(s)/mechanism(s) of failure

f) Occurrence (the likelihood that a specific cause/mechanism will occur—a numerical range of 1 to 10 can be used for probability of occurrence with 10 being almost inevitable and 1 being unlikely)

g) Current design/process controls in terms of the prevention, verification or other activities to assure process adequacy

h) Probability of detection (a numerical range of 1 to 10 can be used for detection probability with 10 meaning that the control will not detect the potential failure and 1 meaning that the control will almost certainly detect the potential cause of failure)

i) Risk priority number (this is the number 1-1000 generated from multiplying the severity, occurrence and detection factors. This element is not essential but does make the result numerical and hence comparable. The higher the result the higher priority needs to be given to it).

Failure prevention features

For each process failure mode ensure that provisions are in place to eliminate, control or reduce the effects of failure. Determine the changes required to make the process robust and update the FMEA in order to show:

❑ Recommended actions, prioritizing action on the highest ranked concerns

❑ Responsibility for actions

❑ Actions taken

❑ Resulting severity, occurrence, detection ranking and risk priority number. This is computed after the actions have been taken so as to indicate the degree of improvement.

Relationship analysis

From the function descriptions and flow charts, relationships should be assessed for conflict of responsibility and authority. These may manifest themselves by:

❑ Duplication of activities and decisions such as checking work more than once, repeating work of other groups, etc.

❑ Decisions taken by personnel other than those responsible for the work

❑ Frequent transfer of product or information between groups within a process

Where such instances appear unjustified, establish the cause and propose arrangements to simplify the process.

Productivity assessment

Actual information flow should be assessed to identify the number of transactions caused by inputs or outputs not being completed before work commences.

The frequency that work recycles the feedback loops should also be assessed to identify ineffective practices.

In developing the ISO 9001:1994 system many more checks, inspections and reviewing points than are actually necessary may have been introduced. Assess decision points and establish whether they are justified in terms of their impact on process objectives and are being taken neither too soon, too frequently, or too late to detect failure.

Information needs analysis

The task analysis will produce a great deal of information about the process and it is necessary to determine what information will be needed for the effective and efficient operation and control of the process. ISO 9000:2000 defines a document as *information and its support medium*. Therefore documents are carriers of information. From the task analysis several pieces of information will have been identified—information that defines or conveys:

❑ Inputs and outputs

❑ Work requirements

❑ Verification requirements

❑ Movement requirements

❑ Methods

- ❑ Reference data
- ❑ Guidance material
- ❑ Records
- ❑ Identity

If the information is not required to operate and control the process effectively and efficiently, it cannot be essential and therefore serious consideration should be given to whether it needs to be documented and maintained. The "document what you do approach" led to a proliferation of procedures and records without a need necessarily being established, other than a perceived requirement within ISO 9000. The needs for documentation should be derived from the needs of the process. From the task analysis a list of information needs can be generated. Having obtained an answer to each question in the task analysis, establish the form in which the information will be conveyed to, within and from the process. A range of information carriers is defined in Table 8.3.

Cultural analysis

If we ask people to describe what it is like to work for a particular organization, they often reply in terms of their feelings and emotions that are their perceptions of the essential atmosphere in the organization. This atmosphere is encompassed by two concepts—culture and climate. Culture evolves and can usually be traced back to the organization's founder. The founder gathers around people of like mind and values and these become the role models for new entrants. Culture has a strong influence on people's behavior but is not easily changed. It is an invisible force that consists of deeply held beliefs, values and assumptions that are so ingrained in the fabric of the organization that many people might not be conscious that they hold them.

Culture is expressed by the values, beliefs and norms that permeate an organization and help shape the behavior of its members. Culture guides an organization in meeting its objectives, in working with one another and in dealing with the stakeholders. Culture is therefore the most powerful force in an organization for making things happen. If the culture is value-based, one can issue instructions, rules and regulations with monotonous regularity with little effect. They will simply be ignored as people go about their work, determining for themselves the right things to do. If the culture is command-and-control-based, expounding principles and values will equally have little effect as people await instructions telling them what to do. If the culture is one that is based on trust, the introduction of approval levels will meet with resistance. If measurement and improvement is not habitual, it will also meet with resistance because it is believed to be someone else's job. Where fire fighting is the norm, people thrive on fixing things and avoid the less macho image of silent study and careful planning to secure success and prevent failure. Things do not change simply because someone in authority releases a document, a policy an edict! There has to be a desire for change, the motivation for change and a leader who will create and nourish the environment in which cultural change will evolve.

There will be cultural factors that are affecting the current performance of a process. Some of these may be advantageous and some detrimental. In moving to a process approach, some beliefs will have to change. The cultural traits that act as drivers and barriers to system effectiveness need to be identified so that the impact of change is fully recognized and accepted by management before proceeding beyond system design. As processes serve to achieve objectives it follows that culture is a key factor in the effectiveness of business processes. A cultural analysis is therefore necessary to establish those aspects of behavior that impact process performance.

Any installation requires a firm foundation and the purpose of analysis to establish what changes have to be made to prepare the foundations for a successful installation. Some of these will pervade all processes and some may be process-specific and not become apparent until much later. What is important at this stage is to identify the key cultural changes that need to pervade the whole organization. By asking the question "What factors affect our ability to accomplish our mission?" or "What factors affect our ability to achieve our objective?" the cultural factors will emerge. Determine what the current perceptions are relative to each factor then look forward to what you want the perceptions to be. The difference between the two is the objective to aim at. By identifying measures and targets for establishing whether the objectives have been met, progress along the journey towards cultural change can be determined. The factors that characterize each of the eight principles of quality management in Tables 3.1 to 3.8 are useful in determining cultural change objectives.

Table 8.3 Information carriers

Information carrier	Purpose
System description	A high level description of the BMS in terms of its purpose, objectives and configuration together with the criteria for evaluating its performance. It may include policies, business process overview and process descriptions and be used for management and training purposes
Policies	Policies are used to define constraints over actions and decisions so that they meet the needs of stakeholders.
Business Process Overview	Diagrammatic description of Business processes that is used to show their interrelationship
Process Descriptions	Descriptions of each business process which show the process purpose, objectives, measures, necessary inputs, flow or sequence of key tasks, outputs, and performance review activities.
Control procedures	Control procedures control work on a product or information as it passes through a process.
Work Instructions	Instructions describe how specific tasks are to be performed or controlled as product, information or service passes through a process.
Standards	Standards define acceptance criteria for judging the quality of an activity, a document, a result, a product or a service.
Specifications	Specifications are used to define requirements for a task, a product, a service or a process.

Information carrier	Purpose
Guides	Guides aid decision-making and conduct of activities.
Blank forms	Blank forms are used to collect and transmit information for analysis or approval.
Blank labels	Labels identify product status and are often disposed of when the status changes.
Notices	Notices alert staff to regulations that must be followed, to precautions and to potential dangers.
Job or role descriptions	Job or role descriptions are used to define the job or role in terms of its purpose and objectives and the responsibility, authority and accountability of personnel together with reference to specifications.
Plans	Plans are used to define provisions made to achieve objectives.
Reports	Reports are used to convey the results of an activity.
Records	Records are used to capture information that is needed for subsequent analysis, decision-making or demonstration.

Resolving concerns

Identify unknowns, conflicts, or differences in approach and resolve with those concerned. Deal with concerns promptly as they may signal more serious problems that will plague the conversion.

System design documentation

The results of system design should bc contained in a process analysis report, a system description and several process descriptions.

Process analysis reports

The Process Analysis Reports should be completed by including the results of all the process analysis activities performed with recommendations as to changes that need to be made. Separate reports may be desirable for business and work processes. (The content is outlined in Chapter 10.)

System description or Management System Manual

A system description is a document that contains the high level information about the BMS. Its life begins following agreement of the Systems Requirements and provides a vehicle for containing the results of the business modeling. It may be consolidated with the process descriptions to form a manual that would have all the characteristics of the old Quality Manual and be a lot more useful. There is no explicit requirement or need for this to be an actual "manual" or in any specific format. It can be, for example a "soft" document held within the organization's computerized information system. Many organizations develop the system using their internal intranets that enable more innovative approaches to developing, maintaining and managing the system.

The system description can perform a vital role in demonstrating that the system has been well designed to achieve the business objectives. It describes the way the business is managed and hence is useful for introducing new people to the organization, for training and for staff development as well as satisfying the requirements of ISO 9001:2000. The contents of a typical system description are itemized in Chapter 10.

Process descriptions

A Process Description is a key document that contains or references everything known about a process. During process analysis, information is collected describing the current process. Following implementation of the recommendations in the process analysis report, the process description is updated to reflect the modified process. Process descriptions may be maintained as discrete documents or combined into a "manual" or other media. The content of a typical process description is outlined in Chapter 10.

Process development plan

As existing documentation is matched to needs and new processes, and tasks and activities are identified, a process development plan should be produced that defines the activities necessary to implement the recommendations from the process analysis. The content of a typical process development plan is outlined in Chapter 10.

Summary

In this chapter *process analysis* has been explained as a number of related tasks performed once the system model has been created and the process development teams formed. The sequence and interaction of tasks and processes is only one element of process analysis, albeit a major part and will include the development of process flow charts but process analysis does not end there.

ISO 9001:2000 requires processes to be identified, their sequence and interaction determined, criteria and methods for their effective operation and control determined and the processes measured, monitored and analyzed. Hence, flow charts are but one aspect of meeting this requirement. If we analyze these requirements in terms of how they have been addressed so far, the results are as shown in Table 8.4.

This chapter has explained process analysis, sufficient for the reader to identify the major changes that need to be made in order to convert an existing QMS.

The next chapter explains how to put the system design into practice.

Table 8.4 How the book addresses the requirements of ISO 9001:2000

Requirement	Addressed by	Chapter
Processes identified	Modeling the business	6
	Process flow charting	8
Process sequence and interaction	System model	6
	Process charts	8
Criteria and methods for effective operation and control	Task analysis	8
	Control analysis	8
	Failure modes and effects analysis	8
	Productivity assessment	8
	Relationship analysis	8
	System requirement deployment	8
	Customer requirement deployment	8
Measure, monitor and analyze processes	Key performance indicators	8
	Current performance assessment	8
	Data recording and collection points	8
	Routing of collected data for analysis	8

Chapter 9

System construction

Learning outcomes

After studying this chapter you should be able to:

❑ understand the steps to be taken to construct a system that fulfills the design criteria

❑ understand why documentation is needed, what needs to be documented and how these needs differ depending on risks

❑ install and commission new processes and change existing practices

❑ understand the importance of behavior in making process effective

Change in direction

ISO 9001:1994 required the system to be documented and the documents to be reviewed and approved prior to issue. ISO 9001:2000 also requires the system to be documented and documents to be approved prior to issue, but the important change arises from the requirement for the processes to be managed. Process management is not simply about creating and managing documents—documents are required but as a means for capturing and conveying information about the processes and their outputs. In order to manage a process one needs to manage all the inherent characteristics of the process in such a manner that the requirements of customers and interested parties are fulfilled by its outputs. This means managing information, resources, behavior and results. The management of processes starts when they are being designed and continues through their implementation into operations and maintenance. There is also the vital requirement for their evaluation and continual improvement. Hence, ISO 9001:2000 introduces a different perspective on system documentation and implementation and presents a significant change in direction as will become apparent from reading this chapter.

Process development

Each business process should proceed through development as defined in the process development plan. At the end of process analysis the components of each process will have been identified to a level where process development can commence. This will invariably involve the following tasks:

❑ Maintaining the process development plan

❑ Producing or revising documentation

❑ Refining the business process flow charts

❑ Procuring additional resources

❑ Redefining people's responsibilities and authority

❑ Designing/selecting data capture tools

❑ Designing/selecting data analysis tools

The output of process development will be documents that describe each of the business processes to a level necessary to ensure repeatable performance providing the people have:

❑ The ability to do the job

❑ The motivation to do the job

❑ The resources to do the job

Providing this environment is the subject of process and system integration.

Reasons for documenting information

To document everything you do would be impractical and of little value. The degree of documentation varies from a simple statement of fact to details of how a specific activity is to be carried out. There are several good reasons for documenting information:

❑ To convey a clarity of purpose and the activities necessary to achieve it

❑ To convey requirements and instructions effectively

❑ To provide a basis for studying existing work practices and identifying opportunities for improvement

❑ To convert solved problems into recorded knowledge so as to avoid having to solve them repeatedly

❑ To provide legitimacy and authority for the actions and decisions needed

❑ To make responsibility clear and to create the conditions of self-control

❑ To provide training and reference material for new and existing staff

❑ To improve communication and to provide consistency and predictability in carrying out repetitive tasks

❑ To provide co-ordination for inter-departmental action

❑ To provide freedom for management and staff to maximize their contribution to the business

❑ To demonstrate after an incident the precautions which were taken or which should have been taken to prevent it or minimize its occurrence

❑ To free the business from reliance on particular individuals for its effectiveness

Reasons for not documenting information

There are also several reasons for not documenting information:

- ❏ If the course of action or sequence of steps cannot be predicted a procedure or plan cannot be written for unforeseen events.

- ❏ If there is no effect on performance by allowing freedom of action or decision, there is no mandate to prescribe the methods to be employed.

- ❏ If it cannot be foreseen that any person might need to take action or make a decision using information from a process, there is no mandate to require the results to be recorded.

- ❏ If the action or decision is intuitive or spontaneous, no manner of documentation will ensure a better performance.

- ❏ If the action or decision needs to be habitual, documentation will be beneficial only in enabling the individual to reach a level of competence.

Size doesn't matter

Size and type of organizations are often thought to influence the degree of documentation needed. However a large organization could be large because of the quantity of assets—2000 offices with 2 people in each. Or it could be large because it employs 6,000 people, 5,500 of whom do the same job. Size in itself, therefore, is not a factor and size without some units of measure is meaningless. Likewise the type of organization will determine what information needs to be recorded but again not the amount of information needed.

Complexity

Complexity is the primary influence on the degree of documentation needed. Complexity is a function of the number of processes and their interconnections in an organization. The more processes, the greater the number of documents.

118

The more interconnections, the greater the detail within those documents. Complexity is also a function of the relationships. The greater the number of relationships, the greater the complexity and channels of communication. Many documents exist simply to communicate information reliably and act as a point of reference should our memory fail us. In the simplest of processes, all the influencing facts can be remembered accurately. As complexity increases, it becomes more difficult to remember all the facts and recall them accurately.

Competency

Competency is a collection of skills, behaviors, attributes and qualifications required to achieve the objectives of the job or role. When personnel are new to a job, they need information, education and training. Documentation is needed to assist in this process for two reasons. Firstly to make the process repeatable and predictable and secondly to provide a memory bank that is more reliable than the human memory. As people learn the job they begin to rely less and less on documentation to the extent that eventually no prescriptive documentation may be used at all to produce the required output.

Competence is the ability to achieve the result required through a demonstrated use of the skills, behaviors, attributes and qualifications for the job. Hence competence may depend upon the availability of documentation— knowing where to locate data essential for setting up a machine, or contacting a customer, or processing an invoice. If the documentation cannot be found, the person is unlikely to be able to do the job and hence cannot demonstrate competence.

Analyzing existing documentation

Once the needs have been identified, the existing information relating to each process can be captured. The Development Teams should analyze the information and match existing information to the need in the process description taking into account the linking provisions.

Linking documents

All documents should have a parent document which invokes its use or creation, therefore a hierarchy of documents should be created which results in there being no document in existence in the organization that is not linked to another in the hierarchy. The principle is illustrated in Figure 9.1. If a document cannot be attributed to a process directly, or through a linked document, there will be no trigger to cause the information to be used. If it is believed that the information would be used without such a trigger from the flow chart, the chart must be incomplete and should be revisited until a need has been proven and linked to the process.

Process installation overview

Process installation is concerned with bringing information, human resources and physical resources together in the right relationship so that all the components are put in place in readiness to commence operation. In many cases process installation will require a cultural change. There is little point in introducing change to people who are not prepared for it. Installing a dynamic process-based system into an environment in which people still believe in an element-based

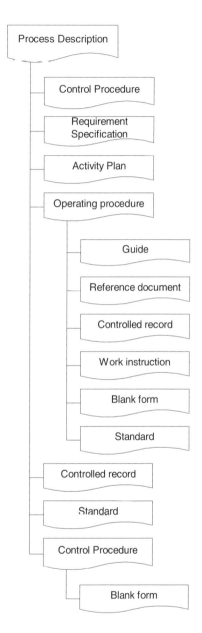

Figure 9.1 Document tree

120

system is doomed to fail. Any installation requires a firm foundation. The cultural analysis will have identified the key changes. Now is the stage during which these changes need to be made.

Preparing the foundations

Top management

To prepare the foundations for a fundamental change from QMS to BMS, the starting point has to be the reorientation of top management. One way of accomplishing this is to get them together and facilitate them to think through the way they operate.

By identifying all the different situations that may arise in a typical business year, get them to assess whether they handle these situations using a functional approach or a process approach. It is not uncommon to find that top management activities feature very few times in the process charts. All the action is at the lower levels. If top management activities are not included in the process flow charts, the processes that achieve the business objectives have not been fully captured.

Staff reorientation

The process development teams should comprise representatives from each contributing function. However, these representatives need to be more than analysts. They need to be the primary means by which staff learn of the changes and the impact they have upon them. General awareness sessions can help but on-the-job instruction is still the best way of making people realize what is involved. Each member of staff will have different perceptions about the organization, its vision, values and processes. Some will be led by others, some will lead the way and some will stand their ground and resist change. Work with the leaders first—they will set examples for others to follow. Leave the "diehards" until last and you may find they come around when they see they stand out in a crowd.

Testing the foundations

In order to test the foundations, use the tables provided in the summaries to each chapter. Alternatively devise another series of typical situations that characterize the difference between the functional approach and the process approach. As process installation and commissioning take place, the change in perception can be tested.

Decommissioning the old way

In some cases new processes can be run alongside old practices but that won't be possible for information-driven processes. Information processes can be terminated while leaving the information dormant in its storage medium. The only change is that people will turn to using new sources of information. It is also wise to retain old software until there is confidence in the new system after which it can safely be removed. A more difficult area to decommission is the informal practice. Few knew it existed at all and even fewer realized it was essential. Those that have been formalized present few problems but those that have been discontinued do present difficulties and are analogous to trying to break a habit. Even when the person is aware of the consequences, the habit persists hence vigilance is necessary until the habit has been eradicated.

Another problem is communication. Some organizations thrive upon communication up and down the hierarchy. Under process management, communication has to be primarily laterally through the process and across departmental boundaries. The extent to which this is a barrier will depend on the prevailing management style. Again it is necessary to be resolute and encourage people to think twice before communicating upwards and ask themselves whether the communication would be more effective if made between the defined stages in the process.

Installing the new processes

The business does not stop while new or modified processes are installed. Installation rarely happens in virgin territory—there are processes operating and delivering results all the time. System installation is therefore carried out in measured steps. In many cases there will be minor modifications to processes such as the introduction of new data recording, collection and analysis routines to measure performance. In other cases there will be new processes to install, such as customer satisfaction measurement. Whatever the case, it is necessary to plan the installation in order to ensure the right people are equipped with the right information and methods to do the right job at the right time.

Armed with the process description, call those concerned together and "walk" through the process. Gain understanding and commitment and secure the resources to make it happen. This act alone may reveal some things that need to be changed in the process construction but it is better to find out at the start rather than later. Installation means also setting up work areas, equipment, new software and databases, communication methods and other aspects of the process. When the physical aspects have been prepared and everyone involved understands their roles, their responsibilities and how they are to act and react, the process is ready for commissioning.

Process commissioning

Process commissioning is concerned with getting all the new processes working following installation. The people will have been through reorientation and will have received all the necessary process information.

Any new resources will have been acquired and deployed and the old practice decommissioned. Installation and commissioning of new processes take place sequentially usually without a break so that current operations are not adversely affected.

During commissioning the following activities will be necessary:

- ❑ Coaching staff in the new practices and testing understanding

- ❑ Testing process flow, interfaces and feedback loops

- ❑ Testing process controls, output quality, delivery and cost

- ❑ Loading databases with process data

- ❑ Testing data retrieval and analysis tools

- ❑ Testing information transfer channels

- ❑   Testing data analysis and reporting mechanisms

- ❑   Testing process improvement mechanisms

- ❑   Testing process robustness, capability and integrity

The old approach was to issue the procedures then walk away. In effectively managed processes, people are led into the job, coached, trained and not left unsupported until both the process owner and the process operator are satisfied that the desired results are being achieved.

Process integration

Process integration is concerned with changing behavior so that people do the right things right without having to be told. The handholding of the commissioning stage can cease. When process integration is complete, the steps within a process become routine, new habits are formed and beliefs strengthened. The way people act and react to certain stimuli becomes predictable and produces results that are required. Improvement does not come about by implementing requirements—it comes about by integrating principles into behavior.

It may take a long time for integration to occur. Habits may not be formed overnight. People need time to practice and while they are practicing they need to be observed so as to alert the process owners to behaviors that need addressing. One way of measuring process integration is through audit. However, the auditor is not only concerned with conformity. The auditor should be looking for actions and reactions to detect whether they reflect the new cultural traits.

Processes operate independently of departmental boundaries and it is at the interface between departments that the greatest problems will arise. The departmental silos protect feudal practices that need to be broken down if the process is to be effective.

System integration

As each process is installed and commissioned the interconnections begin to form into a system. The system will not be effective if the process linkages do not function properly.

- ❏ One process may operate effectively only by causing an interfacing process to run inefficiently.

- ❏ The capacity of one process may become a constraint on the whole system.

- ❏ Restrictive practices in one area may cause bottlenecks elsewhere.

- ❏ The rate of information output from one process may cause overload on the receiving process.

These and many other problems may arise and until all processes work together in harmony and deliver the desired business outputs, the system is not operating effectively. System integration will take considerably longer to accomplish than the three months usually allocated between issuing procedures and the certification audit. System integration has been an option up until now and no auditors have even attempted to delve into the depths of culture and measuring system effectiveness, so a different approach will have to be taken.

Summary

A number of key messages were contained in this chapter:

- ❏ ISO 9001:1994 required systems to be documented and implemented—ISO 9001:2000 requires processes to be managed and this is where there is a significant change in direction.

- ❏ All documents are derived from a process need.

- ❏ The need for documentation is based on process complexity and personnel competency.

- ❏ All documentation should be linked within the process hierarchy.

❑   Process management is about managing information, resources, behavior and results.

❑   Processes will only be effective if people have the ability, resources and motivation to do the job.

❑   If top management activities are not included in the process flow charts, the processes may not be fully captured.

❑   In effectively managed processes, people are led into the job, coached, trained and not left unsupervised—managers don't issue instructions and simply walk away.

❑   Any installation requires a firm foundation—there is little point in introducing change to people who are not prepared for it.

❑   Informal processes are habits that are difficult to break.

❑   The system is not in place until people do the right things right without having to be told and this requires its integration into the fabric of the organization.

Chapter 10

Successful system validation

Learning outcomes

After studying this chapter you should be able to:

- ❑ determine whether the conversion has been successful
- ❑ identify where further improvement is required

Process reviews

To be successful there is an ongoing need to monitor conversion and review the progress made in achieving the defined objectives. In order to monitor achievements it is necessary to set up a mechanism for process reviews. Of critical importance is the measurement and analysis of current organizational performance and the agreement by all regarding how the effectiveness of the conversion will be measured. This will of course be primarily the change in organizational performance.

Timing

At any point in time after commencing the conversion process, set up a review that establishes the status of the system deliverables.

System deliverables

The conversion process generates a number of deliverables that are indicative of the maturity of the conversion. These deliverables include:

System description

This will include:

- ❑ Statement of system purpose
- ❑ Statement of system scope
- ❑ Business objectives
- ❑ System design criteria
- ❑ Context diagram
- ❑ System model
- ❑ Function matrix
- ❑ System performance indicators
- ❑ System performance measurement method
- ❑ System performance results

Process descriptions

These will include:

- ❑ Process objectives
- ❑ Process owner
- ❑ Process inputs and outputs
- ❑ Process flow charts
- ❑ Dependencies (skills, competencies, capabilities)
- ❑ Key performance indicators
- ❑ Performance measurement methods
- ❑ Process performance results

Process Analysis Report

This will include:

- ❏ Current performance metrics
- ❏ Task analysis results
- ❏ Control analysis results
- ❏ System requirements compliance matrix
- ❏ Customer requirements compliance matrix
- ❏ Resource analysis results
- ❏ Process constraints
- ❏ FMEA results
- ❏ Relationship analysis results
- ❏ Productivity assessment results
- ❏ Cultural analysis results
- ❏ Open concerns

Process Development Plan

This will include:

- ❏ New resources to be acquired (space, people, finance facilities)
- ❏ New measuring and monitoring equipment/techniques to be installed
- ❏ Documentation to be produced (information carriers)
- ❏ The target dates for each action
- ❏ The responsibilities for each action
- ❏ The resources necessary to execute the action
- ❏ The outcome from each action
- ❏ The manner in which changes to the process will be implemented to overcome resistance to change
- ❏ The provisions for removing existing controls

129

Status reports

These will address progress of the following relative to the Process Development Plan:

- ❑ Process development
- ❑ Process installation
- ❑ Process commissioning
- ❑ Process integration

Review criteria

Key questions to be addressed include:

- ❑ Have the key processes necessary to deliver the business objectives been identified?

- ❑ Do the key processes accurately reflect how product and information is controlled as it passes between the various parts of the business?

- ❑ Are the definitions and outputs of the processes consistent with the business objectives?

- ❑ Have all external and internal interfaces been accounted for?

- ❑ Do the models show that the controls are compliant with the governing standards?

- ❑ Are the models consistent and coherent?

- ❑ Have processes been put in place for determining:

  - a) Stakeholder expectations?
  - b) Stakeholder satisfaction?
  - c) System effectiveness?

- ❑ Have processes been put in place for managing:

    a) Supplier relationships?

    b) Business information?

- ❑ Recruitment, selection and development of human resources including competence definition and assessment?

- ❑ Physical resources including finances, plant, machinery and facilities?

- ❑ The human and physical work environment?

- ❑ Improvement programs?

- ❑ Would the processes, if implemented as described, enable us to achieve our business objectives?

- ❑ Does everyone in the organization know to which business process he or she actively contributes?

- ❑ Have performance measures for each process been established?

- ❑ Have targets been set for each performance indicator?

- ❑ Is performance being measured against the defined targets?

- ❑ Are the changes to existing practices consistent with the governing requirements and the long-term aims?

- ❑ Have the characteristics for each process been identified?

Results

If satisfactory answers are obtained for all of the above questions, the conversion process is complete. Should any question receive a negative response, further design work is necessary but it may not be a barrier to proceeding with system construction in those areas that are ready. The final arbiter of success is whether performance metrics are showing improvement. The simplest test is to review the trend and be able to explain the change using information generated by the process based BMS.

The following three simple charts should act as an overall reminder that it is not only important to measure performance but to have confidence that the right performance is being measured and that we know why the performance is improving, remaining the same or deteriorating.

Performance

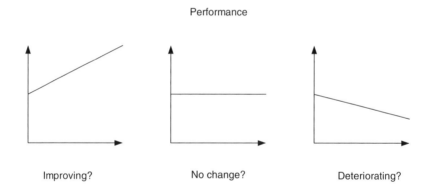

Improving?   No change?   Deteriorating?

The past has only got us to where we are today
... it may not necessarily get us to where we want to be!

Appendices

Introduction

These appendices contain sample flow charts based on typical processes. The charts are not intended to represent any particular type of organization neither do they comprise a complete set of flow charts, The processes are assigned generic names and therefore the processes of any specific organization may or may not include these processes or be attributed the same properties to processes which have been assigned similar names.

The conventions described in Figure 8.2 have been used in laying out the process flow.

The shapes ⬭ indicate an interfacing process or external input

The shapes ▭ indicate a process, task or activity within a process

Only the primary interfaces are shown.

Four levels of charts are presented

Level 0 is the System Model identifying the core business processes and their interfaces.

Level 1 comprises the core business processes and identifies level 2 processes.

Level 2 comprises sub-processes and identifies level 3 processes.

Level 3 comprises the work processes and identifies tasks and/or activities.

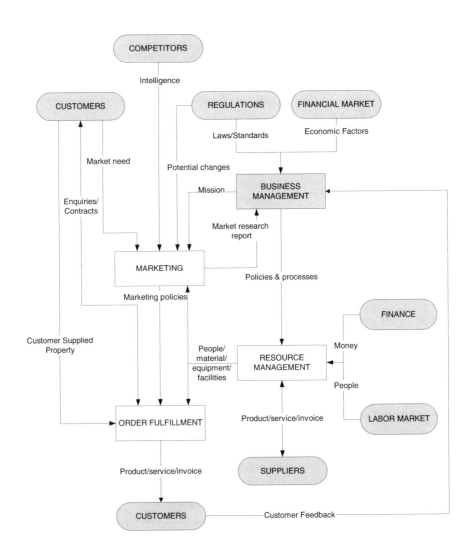

Level 0 The System Model

Appendix A

Sample level 1 flow charts

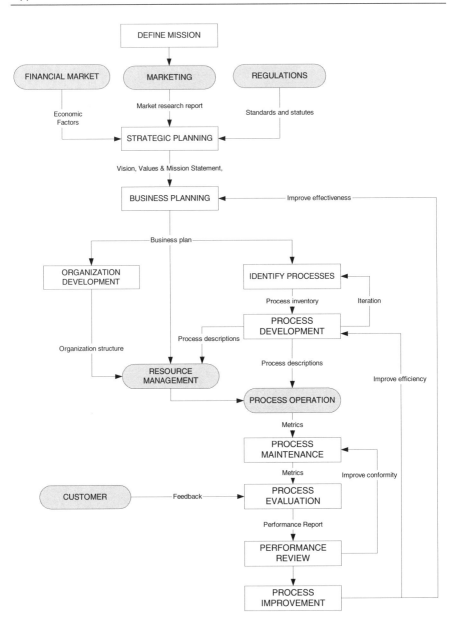

Figure A.1 Business management process

136

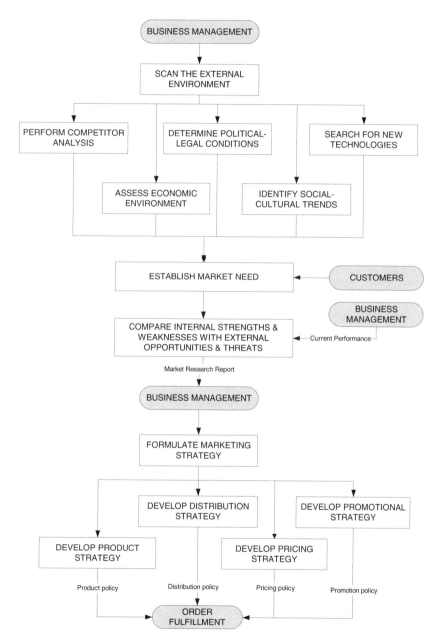

Figure A.2 Marketing process

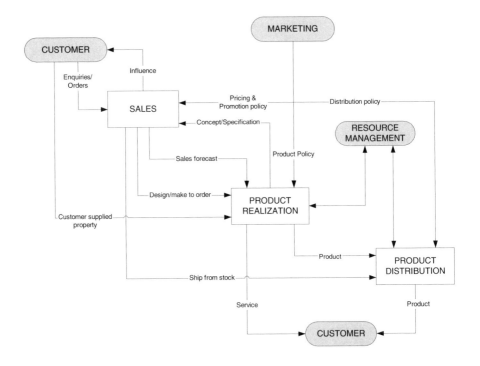

Figure A.3 Order fulfillment process

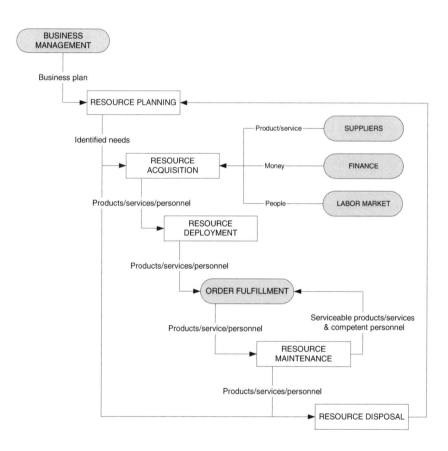

Figure A.4 Resource management process

Appendix B

Sample level 2 flow charts

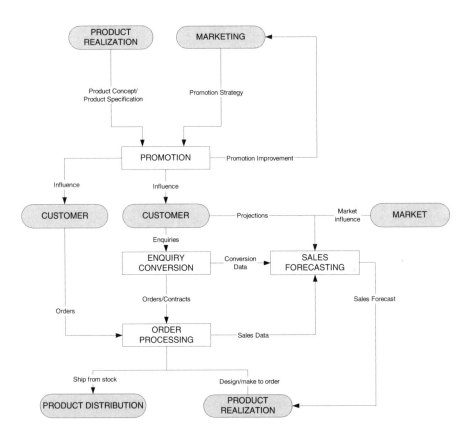

Figure B.1 Sales process

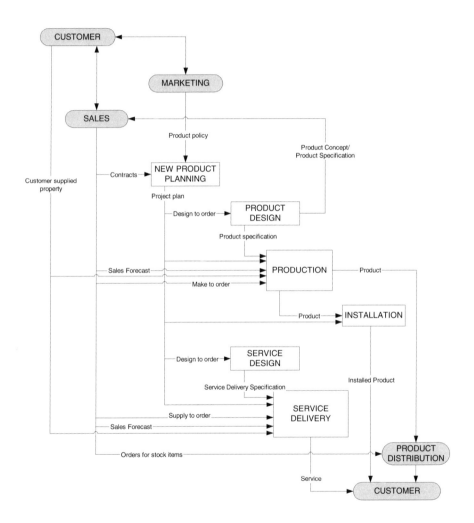

Figure B.2 Product/service realization process

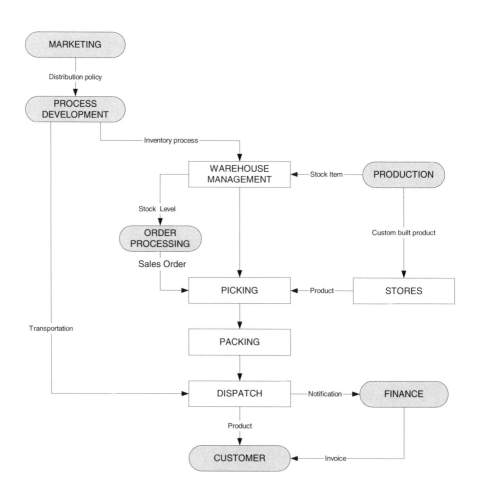

Figure B.3 Product distribution process

144

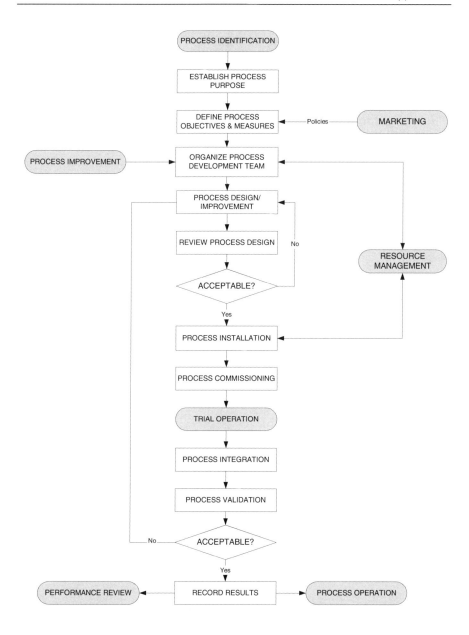

Figure B.4 Process development process

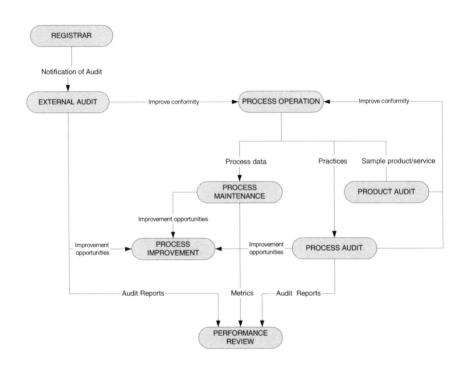

Figure B.5 Process evaluation process

146

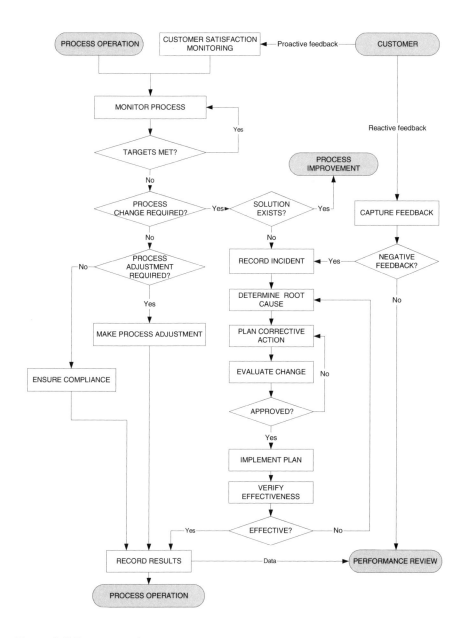

Figure B.6 Process maintenance process

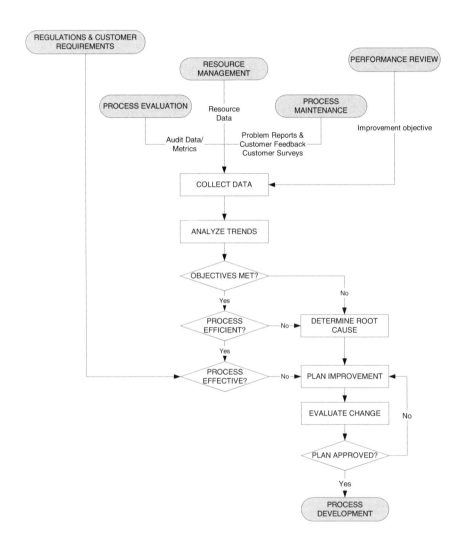

Figure B.7 Process improvement process

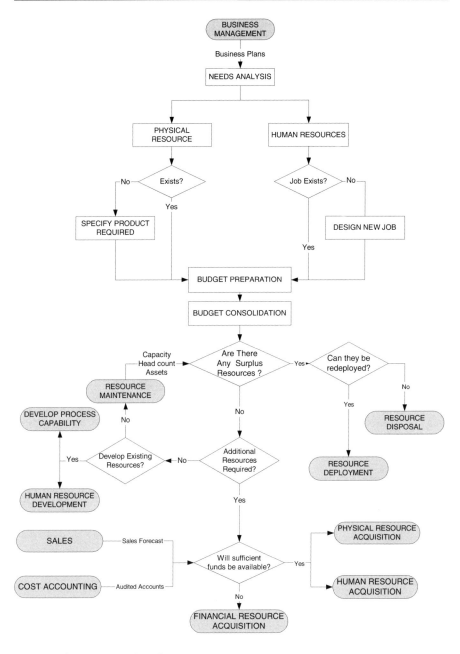

Figure B.8 Resource planning process

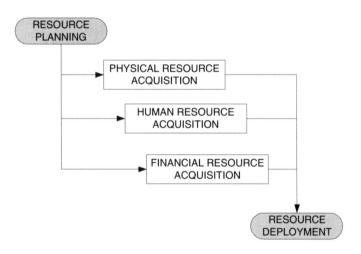

Figure B.9 Resource acquisition process

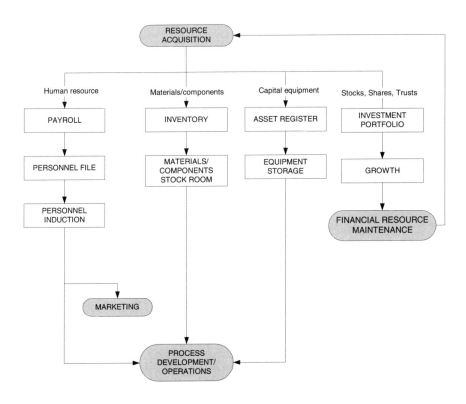

Figure B.10 Resource deployment process

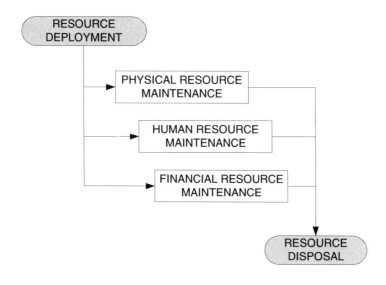

Figure B.11 Resource maintenance process

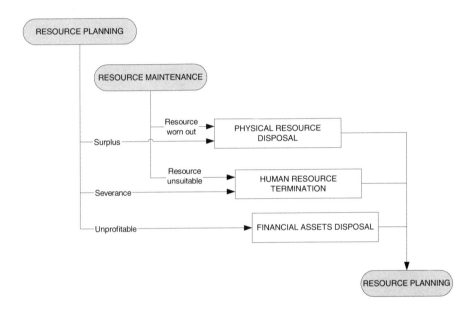

RESOURCE PLANNING

RESOURCE MAINTENANCE

Surplus

Resource
worn out

PHYSICAL RESOURCE
DISPOSAL

Resource
unsuitable

HUMAN RESOURCE
TERMINATION

Severance

Unprofitable

FINANCIAL ASSETS DISPOSAL

RESOURCE PLANNING

Figure B.12 Resource disposal process

Appendix C

Sample level 3 process flow charts

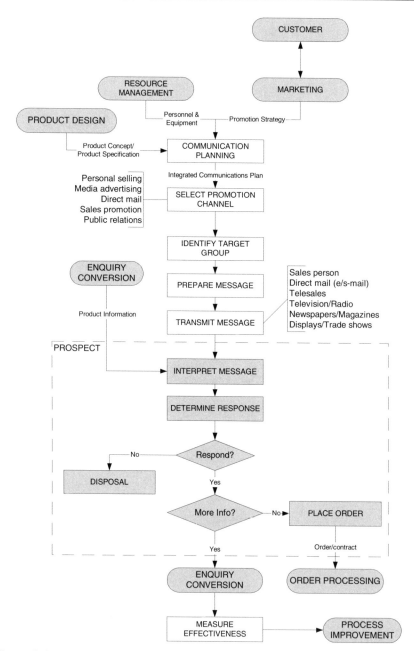

Figure C.1 Promotion process

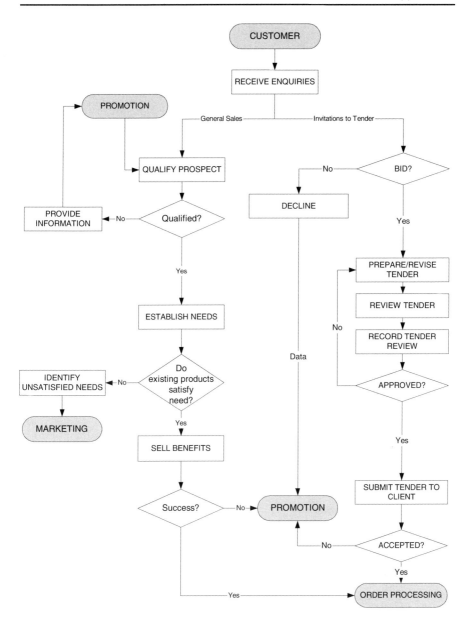

Figure C.2 Enquiry conversion process

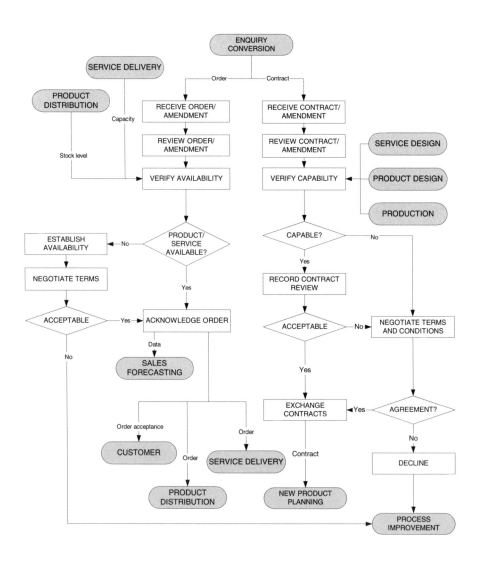

Figure C.3 Order processing process

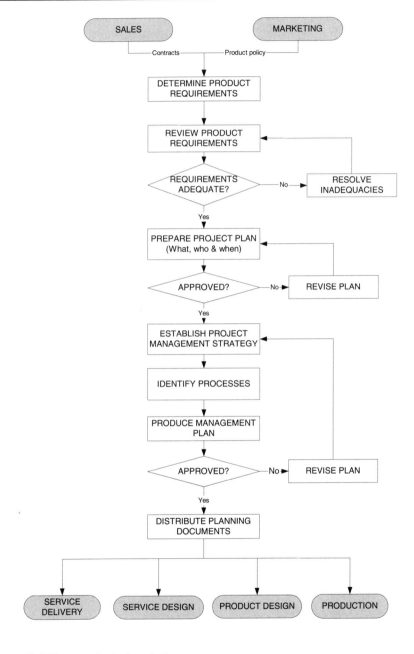

Figure C.4 New product planning process

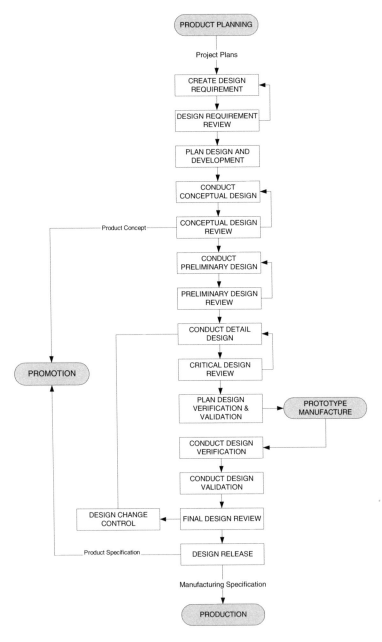

Figure C.5 Product design process

160

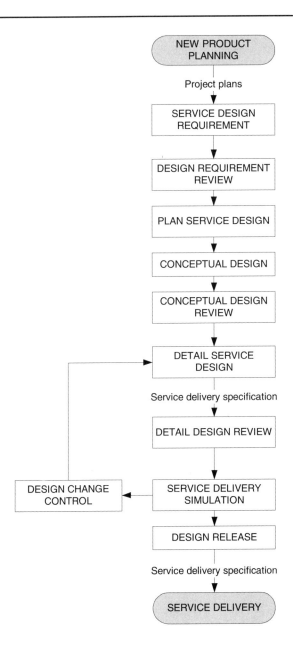

Figure C.6 Service design process

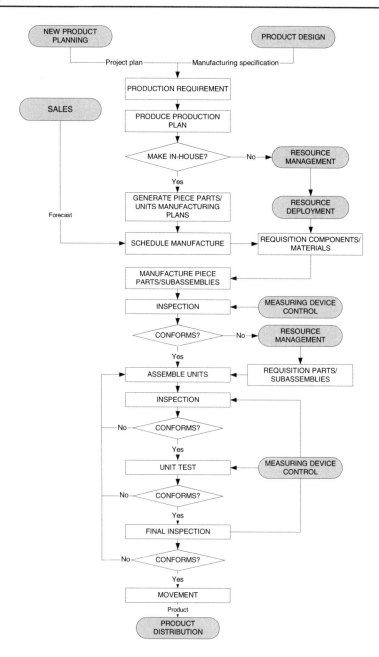

Figure C.7 Product manufacturing process

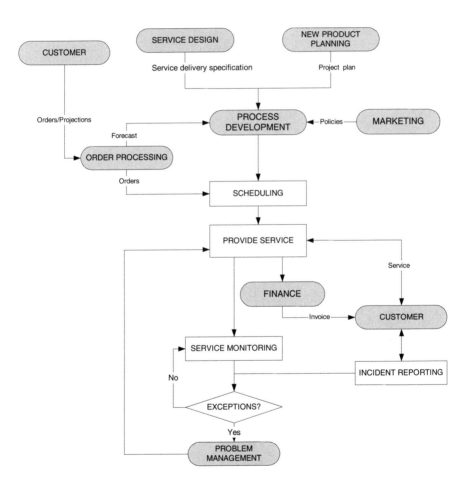

Figure C.8 Service delivery process

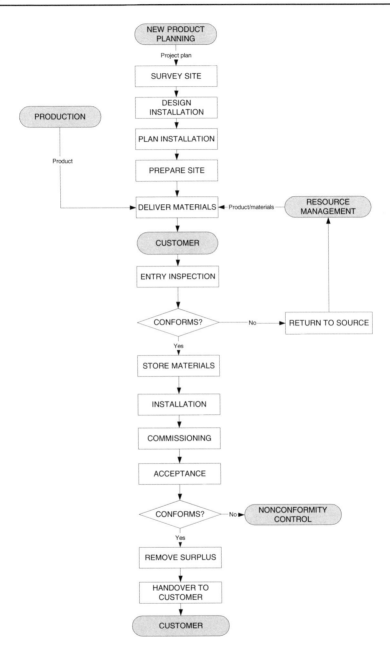

Figure C.9 Installation process

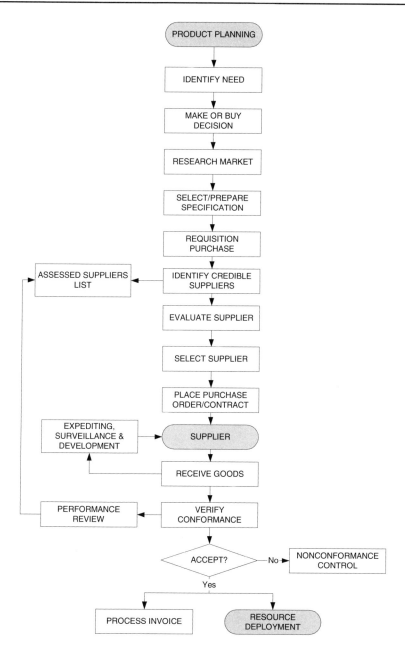

Figure C.10 Physical resource acquisition process

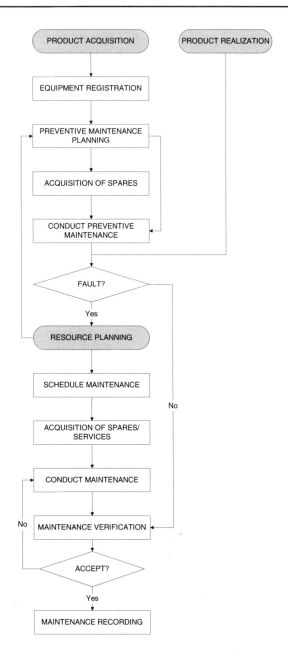

Figure C.11 Physical resource maintenance process

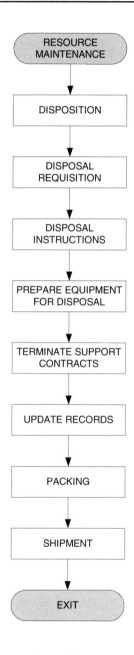

Figure C.12 Physical resource disposal process

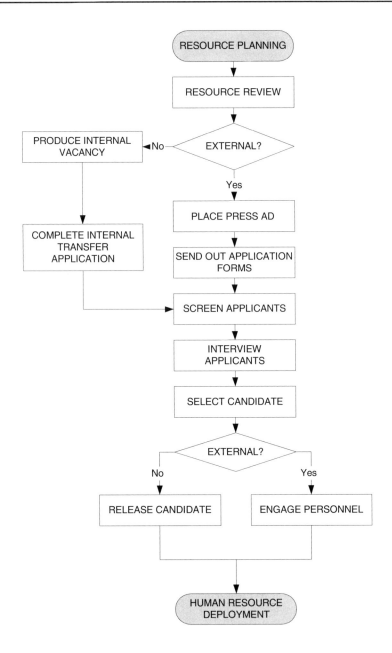

Figure C.13 Human resource acquisition process

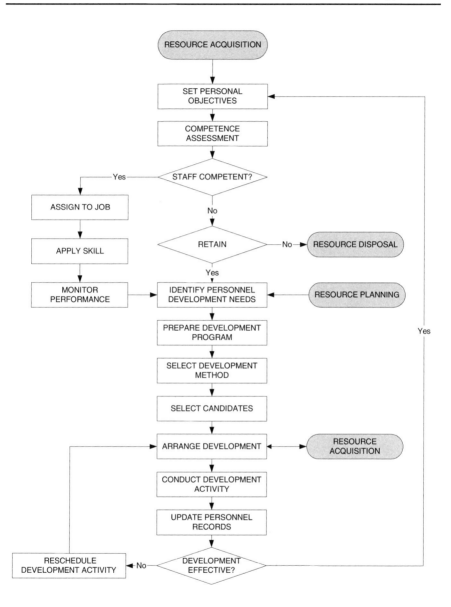

Figure C.14 Human resource development process

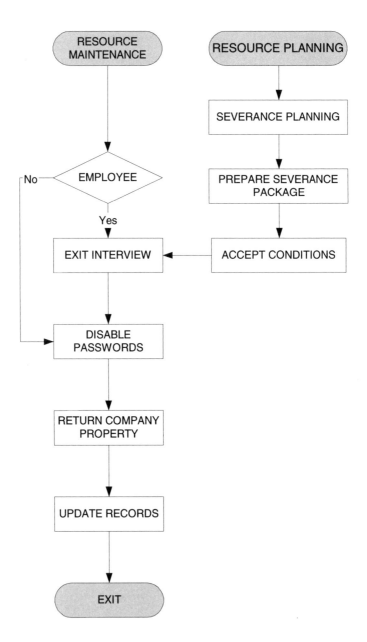

Figure C.15 Human resource termination process

About Transition Support

Transition Support is committed to assisting organizations achieve improved business performance. Our approach to publications is to create a vision of the future and take readers on a journey along which their perceptions will change and by which they will identify new opportunities—in reality a transition that takes the readers from where they are to where they would like to be. Readers will benefit by saving time, gaining knowledge and tools to aid the endeavor.

You can find out more from our web site http://www.transition-support.com

Other publications from Transition Support

- ❑ ISO 9000:2000 Auditor Questions
- ❑ Transition to ISO 9001:2000 ~ Analysis of the differences and implications
- ❑ ISO/TS 16949 Gap Analysis
- ❑ Process approach to auditing

Other services from Transition Support

- ❑ Management consultancy
- ❑ Skills training
- ❑ Technical publishing

Feedback

We would welcome feedback from readers about this book and suggestions for improvement. You can contact us in the following ways:

Tel/Fax: 44 (0)1600 716509 or + 44 (0)1242 525859

E-mail: mail@transition-support.com

Index